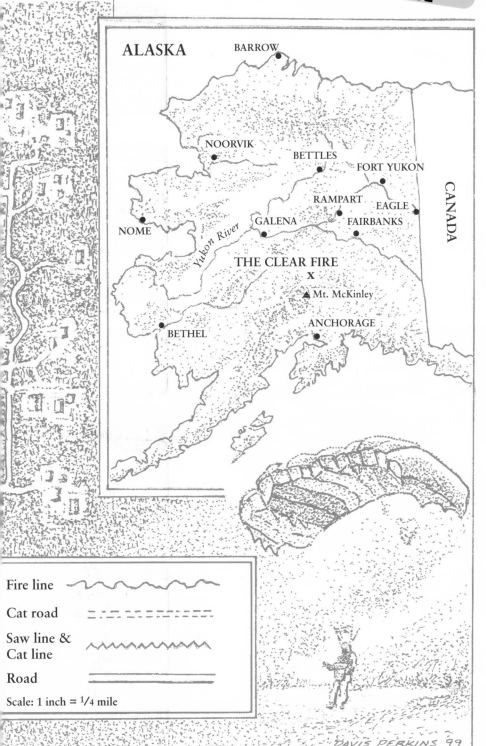

ALASKA

BARROW

NOORVIK

BETTLES

FORT YUKON

CANADA

RAMPART

EAGLE

GALENA

FAIRBANKS

NOME

Yukon River

THE CLEAR FIRE
X

▲ Mt. McKinley

ANCHORAGE

BETHEL

Fire line		
Cat road		
Saw line & Cat line		
Road		

Scale: 1 inch = 1/4 mile

DAVIS PERKINS '99

Jumping Fire

Jumping Fire

A SMOKEJUMPER'S MEMOIR OF FIGHTING WILDFIRE

MURRY A. TAYLOR

Harcourt, Inc.
New York   San Diego   London

Frontispiece illustration by Davis Perkins
courtesy of the Smithsonian Institution.
Endpaper maps by Davis Perkins.

Library of Congress Cataloging-in-Publication Data
Taylor, Murry A.
Jumping fire: a smokejumper's memoir of
fighting wildfire/Murry A. Taylor.—1st ed.
p.   cm.
ISBN 0-15-100589-3
1. Smokejumping—United States—Anecdotes.
2. Smokejumpers—United States—Anecdotes. 3. Taylor, Murry A.
4. Wildfires—United States—Anecdotes. I. Title.
SD421.435 .T39 2000
634.9′618—dc21   99-087608

Designed by Ivan Holmes
Text set in Monotype Sabon
Printed in the United States of America
First edition
A B C D E F G H I J

For the Smokejumpers

What we're all really seeking...is an experience where we can feel the rapture of being alive.

—Joseph Campbell, *The Power of Myth*

"Copy. Jumpers, clear."

"Dispatch, clear."

Meanwhile, the other first-load jumpers and I have run halfway to the plane. The spotter yells, "Hold up. It's not a fire call." Once we stop cussing and kicking at the dirt, we trail back into the shack, mumbling to ourselves.

"That means lunch in ten minutes," Kubichek says.

"Fuck lunch," Mitch grumbles as he picks up a *Penthouse* off the floor. "I'd rather be jumping a fire."

"I'd like to jump this," Quacks says, holding a *Playboy* open to the centerfold. "Wouldn't you?"

"Not me!" Mitch is disgusted now. "This is fire season, and during fire season I think about fires and making money, not women and spending it."

"Man," Quacks sighs. "How'd you like to sleep by her heater?"

"I'll bet she's got a nice personality, too," Troop adds.

Good old Troop, I think to myself. Only Troop could say such a thing and really mean it.

"She reminds me of an old girlfriend of mine," Secret Squirrel says, squinting. "A real barracuda. Had a face like Christmas morning and a heart like Halloween."

"Play it smart," Mitch tells us. "Save your money; stick to chokin' the gopher for now. Once fire season is over, then OK, the women will still be there."

Except for a few scattered noncommittal grunts, Mitch's viewpoint goes largely ignored. The room falls silent while a small group gathers around Quacks to further scrutinize the centerfold.

"Screw it," Kubichek says. "Let's go eat. This is bad for my appetite."

Soon the whole bunch is traipsing by dispatch on the way to the cookhouse, our minds detached—for the moment at least—from thoughts of naked women and wildfire.

Back from lunch, we fell into our usual standby mode. A couple of guys tried a game of horse with the flat basketball. Some returned to the skin magazines. Others, still tired, crawled back under their mosquito nets, trying for a nap. I scanned the standby shack, my eyes drifting to the green magnet board. The right side presented a businesslike listing of the load on standby in Galena plus the names of the other jumpers currently manning fires. The left contained a different list, however. Except for the occasional added name, this list has remained untouched over the years. For those of us who knew those listed, it's a small monument—a roll call of the dead.

The Ghost Load

1. Jimmie B. Pearce . . . . . . took his life, November 1978
2. Bill Resinos . . . . killed jogging—hit by a car 1982
3. "Mouse" Owen . . . . killed skydiving 1985
4. Doug Certain . . . . took his life, March 1987
5. "Granny" Grandquist. . . died—systemic infection—while teaching school, Fort Yukon 1988
6. Dean Johnson . . . . took his life at his home in Redmond, Oregon, March 1988

I looked outside beyond the big bay doors into the gray light of early evening and remembered their happy, carefree faces. Three of them had somehow found it necessary to end their lives by their own hand. The suicide rate is high among smokejumpers. Invariably, it has occurred in the off-season, when we're separated from each other. With their passing, they leave behind a painful subscript to a story cut short. They were all passionate, fun-loving, and spirited. More than that, they were beautiful.

June 14

Yesterday a hard wind blew in from the west. Life in the shack has become not only dusty and dreary but now cold as well. The rest of the jumpers have returned from their fires and were sent on to Fairbanks.

Those of us who remain in Galena huddle in the standby shack, curled up in old chairs, wrapped in sleeping bags and cargo chutes, while the wind rattles the loose metal roofing. The constant tension of having to be fire ready has given way to a strained, restless melancholy. The weather for the next few days is expected to be "stable in between thunderstorm disturbances and generally cooler in the western portion of the state." A few scattered lightning storms are predicted just east of Galena, south of the Yukon River, and below the west end of the Ray Mountains. For the time being, we're stuck where it is no longer hot and dry but cold and soggy.

I crawled up into a cargo chute bin, pulled a chute over me, and tried to go to sleep but couldn't. I lay there thinking about Sally. I'd tried the telephone several times but hadn't been able to reach her. What difference did it make? Mitch was right. If you don't find yourself a girlfriend by June 1, you may as well forget it. Fire season is no time to get romantically involved, and yet adventure has always packed a romantic wallop for me. The sense of losing control of my day-to-day life, blowing from one end of Alaska to the other, marveling at its incredible beauty, plus the stress of standby had combined to increase my yearning for a woman to the point of dire need. Clearly it was time to forget about Sally and whatever refuge her friendship might provide. Yet, letting go has never been one of my strong points, either.

One afternoon, late in my junior year of high school, I was rallying with one of my tennis coaches, Ms. Durban. Suddenly she caught the ball, walked up to the net, and called me over. Ms. Durban knew I was having trouble in school. The rebel-without-a-clue scene—catcalls in the library, setting free a two-foot-long South American caiman in biology, causing a fire in chemistry. At the time, I'd accumulated fifty-three days of after-school detention and was not allowed at any school functions, including assemblies and sports events. During a previous assembly, a near-riot had ensued in the wake of my dropping a scoop of chocolate ice cream off the balcony down Judy Falooti's big-breasted front end.

"Who are you taking to the prom?" Coach Durban asked.

"I'm not going."

"Come over here," she said, pointing to the ground closer to where she stood. I stepped closer. "You *are* going."

"Proms are stupid. I don't want to go."

"Did they say you couldn't?"

"No, but they . . ."

"Then you're going. It's as simple as that. It's your junior-senior prom."

"All the cute girls have already been asked."

"That's nonsense, and you know it," Coach Durban said.

"OK, name one."

Just then, a pretty freshman with long brunette curls came walking out of the girls' gym. We both turned.

"Patsy Liddell," she said.

"Patsy Liddell," I gasped. The possibility of a date with Patsy Liddell was too grand to debate. Even though she was only a freshman, she was already one of the cutest girls in school. Imagine the great consternation it would cause those who thought me a villain to see Patsy Liddell on my arm.

"I can't," I said. "Her father's the head of the Board of Trustees, and they're voting tomorrow whether or not to suspend me from school."

"What've you got to lose? All she can say is no."

I decided that the woman who had coached me so successfully in tennis might be right about a few other matters as well.

That evening, I called the Liddell home and asked Patsy if she'd go with me to the prom. She had to check with her family, she said. She hadn't yet been allowed to date.

I found out later that her father was incredulous. He was considering whether or not I should be suspended from school, not whether or not I should date his daughter. Mrs. Liddell thought it a fine idea. It meant a lot for a freshman girl to be invited to the prom. Her brother Andy, appalled that a monster like me had had the nerve to call their house, threw a fit and stomped out the back door. Her brother Jack, home on leave from Fort Bragg, North Carolina, where he was serving as a member of the 101st Airborne Rangers, took the floor of the family room with the demeanor of a statesman. Dressed in full uniform, he chided them for being too conservative, telling them that it was absurd to not let her go—a chaperoned event at that. Mrs. Liddell and Jack's opinions held sway, and Patsy's answer was yes. And that, beyond all doubt, was a major turning point in my life.

Within two months my behavior had changed. Patsy Liddell and I had struck up a high school romance. I was no longer a clueless rebel but a young man paying attention to something new. The Liddells were a well-to-do family that lived in a sprawling ranch-style home, surrounded by stately eucalyptus trees and rose gardens, situated high on a bluff overlooking the San Joaquin River.

Mr. and Mrs. Liddell seemed to like me and treated me in a way I wasn't accustomed to. That summer Andy and I became like brothers, chasing jackrabbits at night with Mr. Liddell's Jeep. Mrs. Liddell made appointments for me with their family dentist—appointments she knew I couldn't afford, so she graciously hid the bills among her family's bills. She spent hours with me sitting at their kitchen table exploring ideas about life and life goals. Near the end of my senior year, we were in the midst of one of our discussions.

"Where will you be going to college?"

"College? I don't think I'll be going."

"That's a shame," she said. "You're the kind that should. You've far too much to offer this world."

My own family never seriously considered me college material, probably because of my low grades. I hated the indoor nature of school and lived for that day I would be free of it. But even more, I came from the class of people that generally viewed college as something beyond reach.

My mother's father, John Henry Rooney, left Ireland as a young man, fought the Dutch during the Boer War in Africa, then sailed with the English navy around the world. In 1904 he jumped ship in San

Diego and traveled north to Fresno in a buckboard with his Los Angeles bride, Margarita Swartz. The two had six kids by 1919, and he supported them during the lean years of the Great Depression working as a roofer. By 1930 John Rooney was a self-proclaimed Bolshevist and labor organizer. He and my grandmother remained poor until they died, within two months of each other, in the early 1960s.

On my father's side, my great-grandfather, Francis Marion Taylor, was born March 20, 1845, in Ouchita County, Arkansas. He served in the Union Army during the Civil War in Company A, First Regiment of Arkansas Infantry Volunteers. After the war he drifted west, married Sarah Elizabeth Cantrell, a Choctaw Indian, in Henderson County, Texas, and homesteaded 160 acres north of the present town of Novice. In the early 1800s, Francis Taylor's kin had been aristocrats in Virginia and North Carolina, but the war ended all that, killing many of the men, splintering families, and leaving them scattered and destitute.

By the end of 1879, Francis and Sarah had begotten six kids; then Francis died from disabilities incurred in the war. Two of his brothers had served as infantrymen for the Confederacy, and they took up the task of looking after the family he left behind. At the time of my great-grandfather's death, my grandfather, as a five-year-old boy, was left to help his mother provide. My grandfather worked as a cowboy, driving cattle until 1906, when he married. My father was born in 1917 in Coleman County, Texas, forty miles southeast of Abilene. He was the last of the six children of Marion Henry and Emma Francis Taylor. By 1915 they owned two hundred head of registered Hereford cattle and ran a reasonably prosperous ranch. When a drought hit west Texas, the grass died, and they were forced to sell out and give up their land. Moving to California six months after my father was born, my grandfather rode in a boxcar with two mules, a couple of milk cows, a pregnant sow, some farm implements, and the family dog. My grandmother rode up front in a Pullman car with the five older children and my father at her breast.

For the next twelve years they worked every day from dawn until dark and acquired a moderately successful dairy on the west side of the

San Joaquin Valley, near the small town of Tranquility. In 1929 the Great Depression hit and everything they had gained from ten years of grueling hard labor was lost. They spent the final thirty years of their lives on a twenty-acre farm, working it with mules, struggling to make ends meet.

Survival had been the focus of both my mother's and father's families, not higher education. At the insistence of the Liddells, however, I enrolled at Fresno Community College and spent the entire first year taking remedial classes to make up for all I'd missed in high school. I studied hard and my grades were good. My second year I transferred to California State Polytechnic at San Luis Obispo so I could go to school with Patsy's brother Andy. On weekends I would drive back to Fresno, work days for Mr. Liddell, and spend evenings with Patsy curled up on the couch in the family room, listening to jazz and making out.

After my year at Cal Poly, I transferred to Humboldt State College where I eventually earned a bachelor of science degree in forest management. My years in college became the most critically formative of my life. Unfortunately, they were the years that separated Patsy and me. For five years we had courted, planned to marry, and somehow, due to a deep and abiding respect for her parents, had managed to abstain from sexual intercourse.

During my junior year at Humboldt, Patsy wrote and told me that she had fallen in love with a fraternity man she'd met at the University of the Pacific. I immediately drove home, five hundred miles nonstop, and pulled into the driveway of her parents' home, just in time for my headlights to catch her and her new beau kissing good night in the rose garden. Thirty years would pass before I would see her again. Still happily married to the same man, she had raised two handsome grown boys, and was looking just as beautiful as ever.

9

Finally, after five days of mind-numbing standby out west, the Galena siren sounded. Skin magazines fluttered into the air like startled pigeons, chairs flipped over backward, and the first load piled out of the standby shack in a mad dash for *Jumpship 17* on our way to the Clear fire, twenty miles north of Denali National Park.

Tugging and wrestling with our gear, we suited up amid the roar of twin turbines and climbed aboard just as the two Pratt & Whitneys hit full rpm. *Jump 17* winged out over the end of the runway, banked southeast, and flew up the Yukon River toward Yuki Mountain. Relaxed on the floor, I leaned my head against the wall and began thinking fire boss thoughts: time of day, time of year, other resources available, experience of the people on board, recent weather, air tanker availability. On board we had three thousand feet of hose, two pumps, twelve gallons of fuel to run our pumps, and chain saws. I put in my earplugs, grabbed a cargo chute for a pillow, and made ready to get some sleep. Catching a nap on the way to a fire is important, since we never know how long it will be until we can sleep again. A note came back from the cockpit. Each jumper read it in turn. Kubichek passed it to me.

Fire boss—Taylor—the Clear fire—five miles E. of Clear. Smoke seen from highway—homes in the area—one air tanker responding from Fairbanks

Having been seen from five miles away, it could be anything— maybe a homesteader burning debris. I tucked the note into my PG bag and lay my head back again. A half hour later I was jostled awake by Kubichek handing me another note.

Clear fire—State of Alaska—5 acres—1 tanker dropping—1 more en route from Fairbanks—Jump 07 responding from Palmer

The Palmer load would probably beat us there, in which case the first man on their load would be the fire boss. Clear was known for its afternoon winds, being so close to the Alaska Range. Ten minutes later, Kubichek handed me another note.

Estimated 200 acres—100% active perimeter—rolling—black spruce—state troopers are contacting locals. State Div. of Forestry Rep. is Schmole—in helicopter 801 Kilo Alpha— Channel 9. Go ahead and get ready.

We completed our buddy checks, then settled back to wait. Shortly, there was a commotion up front. Two jumpers got to their feet and looked forward through the windshield. Straight ahead about twenty miles, a giant cauliflower-shaped smoke column loomed twenty thousand feet into a cloudless sky.

"You can pick your friends, and you can pick your nose," Quacks laughed, rolling his eyes, "but you can't pick your fires."

As we pulled in over the fire, our spotter, Dalan Romero, stashed his maps and clipboard in the cockpit and climbed back over the jumpers and cargo, shaking his head and shooting me a quick you-poor-devils grin. Dalan's levity soon gave way to serious concern. He rested his hand on my shoulder and yelled over the engine noise.

"When you get on the ground, radio Schmole. I lost contact with him. *Jump 07* got here just before we did, so we'll swing into the pattern behind them and drop tandem. Also, I think you may be getting a third load out of Fairbanks. There are homes in the area—sounds like one of them is already lost." He stopped and stared at me a second, turned something over in his mind, then cupped his hand again to my ear.

"There are people down there—that could make it tough. Maybe you can deal through Schmole with the state troopers to get them evacuated. I think it's five hundred acres by now. According to the helicopter pilot, the winds are erratic near the fire. Be careful on this one, Murry. Keep the crew heads-up."

Dalan snapped the line into his spotter's harness pigtail, braced himself, grabbed the door, and twisted the handles.

"Guard your reserves," he yelled.

Hinging inward toward the tail, the door opened to a great rush of cold air laced with the scent of woodsmoke.

We circled the Clear fire, a typical Alaskan "gobbler," only this time there were people and homes to worry about. The rest of the jumpers crowded the windows, scrutinizing the scene for any details that might later prove useful in planning tactics.

I looked down through the gaping door of *Jumpship 17* at three thousand feet of a raging fire front. Flames across the head were sixty to eighty feet high. Some in the interior appeared larger—red balls of igniting gases ripped free from the earth and rolled up, trailing black coils of smoke.

Jump 07 set up in the drop pattern and we fell in behind it. On our initial pass, Dalan dropped the first set of drift streamers, and we watched them wave one way then the other in the turbulent winds. I pulled on my helmet, snapped the chin strap, dropped the mask down over my face, and put on my Nomex fire-resistant jump gloves. Apparently satisfied with what he'd seen, Dalan turned to me, holding up two fingers. Kubichek and I hooked our drogue static lines to the over-head static line extenders and moved into the exit position.

"Looks like about 450 yards of drift. The wind is strongest down low. Stay wide of the fire. The jump spot is at the tail where those two roads cross. See it?" I nodded yes. Dalan's head swung back out the door. Within seconds he turned back.

"Get in the door," he yelled. I dropped into the door, braced myself, and performed my four-point check.

The Kantishna Plain lay below in a flat incline that ran east into the dark green outer range, the Toklat River Valley, the Wyoming Hills, and on into the heart of Denali National Park.

Dalan pulled his head out of the slipstream and lifted his arm.

"Get ready!"

The slap came down on my shoulder, and in the next instant I was out and falling below the belly of *Jump 17*, counting in my head and keeping my eyes on the horizon. Once stable, I pulled my rip cord, accelerated for a split second, then felt the tugging reassurance of a canopy beginning to open. High above a burning landscape, I watched my chute billow out in perfect form. Grabbing my steering toggles, I turned to face one square mile of violent flames whipping back and forth and a smoke column that cast an ominous shadow far out over the land.

Several rough dirt roads crisscrossed the area near the jump spot. I saw the homes. One had a flame front moving directly toward it. Another stood off to the side, clear for the time being. A third had burned to the ground.

Kubichek and I held two hundred yards wide of the fire and came flying into the jump spot just as the ground wind shifted. Quacks got in too close, blew out over the fire, disappeared in the smoke, and crashed down twenty yards inside the fire. Three of us ran over and pulled him, cussing and coughing, out of the hot ashes. His parachute had a few burn holes, but the heat wasn't intense enough to cause him serious harm. About the time the last jumpers from the first two loads landed, two other jump ships arrived and began throwing streamers.

"Nice job out there, Quacko," Mitch yelled. "I sort of liked that spot over in the fire, too, but decided that over here in the cool, green grass along the road might be just as good."

"Wait'll Ed gets a look at that parachute, wonder boy," Fergy yelled. Ed Strong was the parachute loft foreman. "You'll be on two lists—the jump list and Uncle Ed's shit list."

Dalan's brother Rene had jumped first from *Jump 07,* so he would be the jumper in charge. When I met with him, he was standing on top of an upside-down wrecked car talking on the radio, trying to get Schmole to agree on what the line authority would be. Since the fire was burning on state-protected lands, they decided Schmole would remain fire boss and Rene would act as line boss. Rene would direct the initial attack on the ground and Schmole would coordinate between State Forestry and BLM, both on the fire and back in Fairbanks.

Rene immediately called for a plans briefing. While the other jumpers assembled our paracargo, Rene, Troop, and I knelt down in the road and began drawing in the dirt with sticks. Our first priority was to provide for the safety of the local people and save the two homes. Rene assigned Troop the responsibility for the left flank and me the right. We would work together to secure the tail of the fire near our jump spot, then move up each flank in hope of rounding the head of the fire before the hottest part of the following day.

"The way it looked from the air," I said, "we could do a lot of good up the right flank with a burnout from this road, plus the one that runs east-west, and then north toward the head." I tapped my stick on the ground and searched their faces for a response.

"Go ahead, Murry," Rene said. "Give it a try. That'll be your side."

"We'll try to flank it to the Nenana River. It's going to hit a pretty big stand of birch just before it gets there. From the air, it looked like that could slow it down enough to keep it from spotting across."

"Troop," Rene said, tapping the toe of Troop's boot with his stick, "the houses are here and here. When the air tankers arrive, you'll have first priority. After the houses are secured, we'll play it by ear."

I turned to Rene. "Have they said anything about what fire engines, dozers, or manpower we can expect in here by morning?"

"That's all still up in the air. I've had a hard time getting Schmole on the radio. But he did tell me that they've made this a priority fire. The air force has a communications facility right out in front of it, just across the river.

"We can catch this with the forty jumpers we've got," Rene went on. "But Schmole is thinking of ordering a Class II team. That'd waste too much money. Once this thing lays down, all it's going to need is a few crews. I asked him to hold off ordering the team until three tomorrow morning. If we're doing well enough, he's agreed to reconsider. We can catch this fire! *He* doesn't believe it, but *we* do! Murry, I'll get up in the air when he gets back with the helicopter. If you can, try to be up on the head at least by two o'clock. I'll need a firsthand report by then."

"Well, let's do it," Troop said, eager to do battle. "The longer we sit here gabbing, the bigger the dragon gets."

As bad as it looked, we all believed we could catch the fire. It was still relatively damp deep down in the tundra mat, and the ground fuels (leaves, twigs, branches, and grass) would absorb some of that moisture during the night. Such is usually the case in mid-June. Knowing the moisture level in the tundra mat is critical when anticipating Alaska fire behavior. The nearby snow-filled mountains and the river would also increase the night's humidity.

During the course of our meeting, the other jumpers had assembled our paracargo on the road, stuffed their PG bags with food, and organized the chain saws, fuel mix, bar oil, tool packs, fusees (railroad flares), Pulaskis, and piss pumps for an all-night fire fight.

"All right," Troop said. "That's enough. Let's catch this son of a bitch."

Troop went west around the left flank with nineteen men, and I started right with twenty. Rene stayed along the road near the tail to direct fire trucks and other resources as they arrived. The initial stretch of my side, the right flank, was crisscrossed every quarter of a mile by dirt roads that served the local homesteads. Taking into account the location of the roads relative to the wind, we used the railroad flares that come in our fire packs to start a burnout between the roads and the main fire, extending it northeast to a point about a mile from the jump spot. Two state fire engines arrived and began patrolling the roads for spot fires.

Burnouts differ from backfires in that the main fire's direction is not intended to be influenced by the burnout. You simply burn out the fuels between your line (in this case the road) and the main fire. Backfires, on the other hand, are lit with the intent of creating enough fire to affect the main fire's direction—usually pulling it toward the backfire. Burnouts and backfires can be risky. A favorable wind, one at your back, can carry your lit fire toward the main fire. A favorable wind— or no wind at all—is a fundamental prerequisite to success.

The main fire covered a thousand acres by midnight, two miles long and a mile wide. Troop's jumpers, assisted by some air tanker drops, had managed to save the two homes, but they were having a hard time progressing up the left flank. About that time Air Attack informed us there would be no more air tankers for the night. Pilots require a minimum eight-hour break, so the decision had been made to save them for the critical part of the following day. Troop and his bunch came up on the radio.

"Troop. We're spread too thin," I heard Seiler say. He was panting hard, out of breath. "We've got fire behind us."

"Yeah, that's great," Troop howled back. "We got her now."

"Troop. We've got to slow down," Quacks pleaded. "It's jumped that little slough. We need some guys back here quick."

"Yeah, man! This is beautiful," Troop said. "We got fire all over the place up here, too."

"Troop, this is Robinson. I'm way out in front. I haven't been able to hear our chain saws for over an hour. I think there's at least a mile between us, and I'm still not to the head yet."

"Troop," called Fergy. "That hose we left at the first house—it burned up. I can't find any more. I'll go back to the jump spot and see if there's any there."

"You bet," Troop replied. "You guys are the greatest. Go get 'em. We're making unbelievable progress."

To the jumpers overhearing the conversation it was obvious—Troop had come down with another case of dragon fever. The Don Quixote of smokejumping was once again engaged in mortal combat with this, his latest windmill.

Having recently traveled to China on his own money, Troop had gabbed his way through miles of red tape until he had the ear of the Chinese forest fire leadership. He tried to convince them of the wisdom of using smokejumpers at their remote and destructive forest fires. "Once before, in 1965, we try it," they had told Troop. "We kill too many people."

Troop later explained. "The problem is they just don't relate to the John Wayne stuff over there."

Still Troop pressed on, designing and marketing a Chinese smoke-jumper T-shirt. The proceeds went toward funding the manufacturing of his latest invention, the Universal Wildfire Tool, or "The Dragon Slayer" as he calls it. The T-shirts were white with a design that depicted China in red, outlined by black borders, and a big green, slit-eyed dragon gripping the entire country with all fours and blowing fire out its nose.

"Roger that," he howled. "We'll have this section in no time. You guys are *animals*. Ten-four, good buddy. The night air will help. There's fear in her eyes. We got her by the tail."

Once, in 1988, I saw Troop drive a group of jumpers from the Lower 48 to the edge of madness. We were on a five-hundred-thousand-acre fire just south of the Yukon River, twenty miles east of the Trans-Alaska Pipeline. The fire's perimeter had been estimated at 350 miles. Troop had rallied his charges in pursuit of one flare-up after another. Three weeks into the fire, while walking a broad expansive ridge back to camp, I looked down at the spacious headwaters of Victoria Creek and spied him and his crew, tiny ants working a ten-mile stretch of smoky fireline. When I got closer, I waved.

"Hey, Troop. How's it goin'?" I yelled.

"Well," he shouted, full of cheer and filthy as a warthog. "I'm doing great, but these guys hate my guts."

The crew ignored him. They looked like prisoners in irons, which was pretty accurate since they were stuck with a madman bent on beating out 350 miles of fire line with only burlap sacks stuffed with moss.

"Hell of a fire," I said, waving my hand toward a hundred miles of the Yukon Flats.

"Boy! I'll say," Troop sang out. "But I'll tell ya, Murry. You gotta love these guys. They just won't give up."

Two days later, word came to us from Fairbanks dispatch saying that all the jumpers from the Lower 48 could be released from the fire if they wished. Troop's charges were all asleep in their tents. One of our guys went to ask them if they wanted to leave. "Hell, yes!" they said, jumping up, packing their gear, hurrying to the helispot, and huddling there three hours waiting for the helicopter.

"Damn," Troop said, bewildered, as we watched them fly away. "They never even came to say good-bye."

Around midnight the air cooled and the Clear fire lay down. On the right we had advanced a mile toward the head. The burnout had gone well.

From the jump spot at the south end of the fire, the fire line followed the homestead roads. As we burned out, we moved north, then east, giving the control line the appearance of stairs leading upward to the right. Where the last road turned and ran east, we started cutting chain-saw line north through stands of thick black spruce. By one o'clock, most of that sector had cooled down to flame lengths averaging two to three feet, except for the scattered flare-ups.

I needed to be more certain what lay ahead, so I had my crew take a break and eat. While they tore into their rations, I walked through part of the burn toward the head and came upon an old Cat road. I listened for Troop's chain saws, hoping they would be close, but all I could hear was the crackling and popping of the fire. I kept walking. The fire was apparently bigger than we'd thought.

After a while I turned around, figuring that my weary bunch would be finished eating. This was no time to tempt them with an extended break. I walked the Cat road back through smoke that smelled strongly of pine tar and burning pitch—pungent, acrid, and tangy. Stump holes glowed orange-red as smoke curled up from the peat ground cover in light blue tendrils. The towering gray column of late afternoon had dispersed into a vast smear of pink and orange, a glorious backdrop to the black silhouettes of dead trees. The forest that had grown luxuriant and green for decades now stood naked and dying, the limbs of some trees twisting down to Mother Earth. Others, gnarled and grotesque, seemed to have been struck dead while pleading mercy from a beautiful but indifferent sky.

"Taylor, Romero." I pulled my radio from its holster.

"Romero, Taylor."

"How's it going?"

I gave Rene an update. We'd had a successful and complete burnout, but still were only halfway up our flank.

"OK, sounds good," he said. "There'll be a D-8 . . . dozer in here in an hour or so. Can you use it?"

"I don't know yet. I need to get these guys on some chain-saw line, then I'll finish reconning the head. I'll know then. If Troop needs the Cat, send it his way. If not, hold it there. It's going to take a while to find out if we'd gain anything with a Cat line."

"Copy," Rene said.

I could tell Rene was getting tired. His tongue sounded thick, his words slurred.

"I'll get back with you as soon as I get a better look at the front of this thing. It may be more than we think."

"Sounds good, Murry . . . Keep in touch . . . Romero, clear."

When I got back to my crew, they'd just finished eating.

"What's it like?" Boatner asked, shoving what remained of his dinner back into his PG bag.

"A lot of open line. Maybe a mile and a half. I didn't get all the way."

"This son of a bitch is *big*," Wally Wasser said. "We might be in for an ass kickin' tomorrow."

"We might," I said, taking a long look at him. "One thing's for sure. With the thickness of the black spruce, if we can't hold this thing through tomorrow, she'll burn clear to the park."

It was time for me to make a tough call: work the crew all night or give them a rest. When a fire situation worsens and progress on containment isn't sufficient to warrant an all-night push, the better option is to get some sleep—usually between three and six A.M. With a little sleep, crews are likely to be better prepared for a tough fire fight the following day.

Once the spread is stopped, holding the fire through the next day becomes critical. By the second day—given favorable weather—large fires usually cool down to where they can be held within containment lines until they're totally put out.

When help is on the way, then pushing all night makes better sense. In our case, reinforcements weren't likely to arrive until the following afternoon. Too late.

I looked at my watch—2:15.

"OK, you guys. Here's the deal."

I explained what I'd seen while scouting and related what I'd heard over the radio. I wanted everyone to understand what we were up against, and I needed to hear how they felt about it.

"Maybe we can catch it by tomorrow night. I mean, today night. I mean, tonight night." I stopped.

"Night-night?" Scott Dewitz asked. "Hey boys, it's night-night time for Old Leathersack. The old man's brain's working on a caffeine deficit."

"Later. This is gonna be one of those close ones. If things go our way, we'll be beat up for sure, but we'll have something to be proud of. If they don't . . . we'll just get beat up and not have shit to show for it."

"We'll catch this bitch," Dewitz said.

"Guaranteed," snapped Wally.

Off in the distance, a new column of smoke boiled up out of the woods near the head. We heard the roar, and then ash began falling silently like snow all around us.

"Pretty impressive for this time of night," Boatner muttered.

It was time to make our move. Considering the distance to the river, the thickness of the spruce, and the fatigue of the crew, it seemed an impossible task to cut a saw line that would be wide enough to hold during the heat of the following day. I knew—as did everyone else—that we might just be pissing into the wind.

Regardless of doubts, I went ahead and divided the crew into three groups. Three men were to patrol back to the jump spot and check the area of the burnout. At the jump spot, they were to get food and water and bring it up front. Six men would begin cutting a saw line around a one-acre slopover that had already crossed one of the roads we'd burned out from. The last group of nine I put in the woods where we had finished the burnout. They were to punch saw line toward the head. In ten hours, we needed to be tied in with Troop.

I left them to sort out the details of their assignments and took off again toward the head. Through a silhouette of charred trees, I watched as the sky in the north brightened. The sun had set at 12:50 and would rise about 2:55. Somewhere out there, below the horizon, I knew it had begun its upward arc.

After a mile I came to what seemed to be a corner as the fire's edge turned left. I turned and followed it another quarter mile, stopping to listen now and then, but I still couldn't hear Troop's saws. Maybe they were taking a break. Surely they couldn't be that far. I was tired. My knees ached the way they do when a body needs sleep. Approaching the river, I could see that the fire had run straight into a large stand of birch. With the night air, cooler temperatures, and no wind, the head had laid down. The flare-up we'd seen must have been an island of green in the interior. There were no big flames along this part of the head, just a few creeping fingers and heavy smoke.

"Troop, Taylor. Channel 9."

"Yeah! Go ahead, Murry." Troop was out of breath.

"Have you been to the head yet?"

"No, but I sent Robinson up there, and he says she's laid down real good."

"It has for now. How do you feel about holding your side tomorrow?"

"Oh," he assured me, "no problem, I got some real animals over here. We'll do fine . . . unless we get a wind."

"Uh-huh. That'll probably be the hot side tomorrow . . . I mean today. We'll see how it goes. How far do you have to go with your saw line?"

"Maybe a quarter mile. Then it hits birch."

"Good job, Troop. You need that dozer?"

"Negative! We might need it if she heats up later, but not now."

"Copy. Thanks, Trooper. Talk to you later. Taylor, clear."

"Animals, clear."

Good old Troop. The mighty warrior. One of the great smoke-jumping legends—the man who had jumped more fires than any other jumper in history, leading his men in the smoke and devastation, giving orders, encouraging them, fighting a wildfire.

As a young marine lieutenant, Troop had survived two 2-year tours of active combat in Vietnam and had been decorated with two Bronze Stars and a Purple Heart. I once wrote to him asking about his medals.

"In regard to your question about medals," he wrote back. "I got a few but mostly for the wrong reasons. What I'm most proud of is just having the honor to have served with my marines. No matter what anyone's view of the war was, they would have liked the men I fought with. They had manners, they took care of each other, rescued the wounded, suffered with dignity, and accepted that they would die in

Vietnam. The medals don't mean shit. They're lost in a drawer somewhere, and some fine young men never came home. My greatest honor is to have known them."

I radioed Rene and told him that the head looked good—that it had run into some birch and damp grassy country near the river and that we wouldn't need that Class II team, just a half-dozen good crews—especially the hotshots from Fairbanks, if they weren't already on the way.

"Sounds good," Rene said. "I like it. But I think Schmole's going to go for the team anyway. Too risky, he says. With it being a state fire, it's his call. Anyway, we got one hotshot crew coming and five EFF crews, but they won't get here until noon tomorrow . . . ugh . . . today."

"Rene, where's that D-8?"

"It's right here. You want it? Seiler's the swamper, take him, too."

"I'll take them both! Send 'em this way. Tell 'em to meet me where the right flank leaves the road."

"Will do. They're on their way."

I looked at my watch—3:05. I pulled two instant coffees out of my shirt pocket, tore them open, and poured the contents into my left hand. I snapped off the lid to my canteen, popped the freeze-dried coffee into my mouth, and poured in some water. I sloshed the mixture around, swallowed it, and then walked back to where the guys were working on the saw lines.

I met with Boatner awhile, then walked to the main road to wait for Seiler and the Cat. The guys cutting saw line around the slopover had almost completed their assignment. I told them to join the others working toward the head when they finished.

Down the road, a fire engine backed up, its red lights floating in clouds of dust and drift smoke. To the south, the highest peaks of the Alaska Range were gathering light.

The caffeine kicked in and my guts ached something terrible. I took out a Snickers bar, ate half, and stuffed the rest back in my shirt.

"Shit!" I muttered to myself.

I felt rough and strung out. I thought of better nights in warm beds with flowered pillowcases and women with soft breasts. I wondered how long it would be until the Cat showed up. I needed a nap.

Then I felt the coffee sizzling in my belly. *Maybe a campfire,* I thought. At the edge of the road, I found a hot stump hole, tossed in some sticks, dragged up a charred log, sat down, and stared into the fire. It was warm, but the heat made me sleepy, so I stood up and jumped up and down. Then I sat down again, mortified that the guys in the fire truck might have seen me.

In the distance the Cat came clattering up the road, banging and screeching along, its tracks hammering, its engine throbbing inside its great body. Two headlights flashed around a turn, and through the smoke and dust the beams came at me, crisscrossed at odd angles. I stood up, kicked dirt into the stump hole, took a drink of water, and prepared to greet eighty thousand pounds of earth-shaking, turbo-charged, tortured steel bolted to a sixteen-foot dozer blade. The Cat screeched to a stop in front of me, dust boiling everywhere. I climbed on, shook the skinner's hand, nodded to Seiler, and took a seat on the toolbox.

"Straight ahead," I said, pointing to where the Cat road led to the burn. Instantly the D-8 lurched ahead, its roaring and clattering a tribute to our perverse adoration of mechanical mayhem.

For the next twenty minutes we roared through the burn, choking on dust, pitching over fallen trees, slamming and clanging in and out of stump holes. Ash lifted by the tracks swirled up and was blown forward by the Cat's big fan, creating a storm of dust and smoke and diesel exhaust. We pursed our lips and squinted blindly, but the Cat did not slow down. Seiler and the skinner looked like extras from the movie *Night of the Living Dead.*

The sun cleared the horizon. 4:15 A.M. We were eight hours from the heat of the day. Progress on the saw line had been steady but too slow.

"This is it!" I yelled, turning to the skinner. He cut the throttle, jerked the clutch, and the tractor halted abruptly. We climbed down and walked off a ways to get clear of the noise. The Cat chugged obediently in the settling dust as the skinner pulled out a pack of cigarettes, shook one into his mouth, and lit it.

"From here on," I said, "I think the birch will hold the fire until we get more help. It may burn some, but she's run herself into a corner."

Seiler and the skinner nodded in agreement.

"How much fuel you got?"

The skinner pulled off his baseball cap and scratched his head. "Oh, I'd say around five hours' worth in the Cat here, but there's more back at the low-bed."

"Good. Here's the deal. Boats and the rest of the guys are building saw line this way, but they'll never make it. We'll tie in here and head back toward them. Al and I'll recon. We'll try to keep as close as possible to the edge of the burn, but we'll have to leave some fairly big islands of green. It can't be helped, the fire's edge is too crooked to go direct. It'll take a mile of Cat line to tie into Boats's bunch, but at least we'll have something to make a stand on. It's our only chance to hold this flank."

"You been out through there?" Seiler asked.

"Not all of it, but enough to know it's dog-hair thick. It'll be fast going for the Cat, so you'll have to watch out to keep from gettin' run over."

We agreed I'd do the lead scouting, sixty to seventy feet out in front. Al would follow me, and the Cat would come after him. Since the woods were too thick for the operator to see more than thirty or forty feet ahead, this put Al in constant danger.

"Keep your eye on me," the skinner told Al. "I get involved with what I'm doin'. I can't always take time to watch."

The skinner stepped behind his Cat and took a leak.

Seiler looked at me and shook his head. He was covered with ash, his teeth and eyes looking like white marbles stuck on a tar baby.

"You know that house on Troop's side, closest to the fire?" he asked. "Boy, were those people glad to see us. We beat flames for over an hour, then the air tankers came in and saved the day. I tell you, those people were scared, and so were we. They were so happy that we saved their house, they started crying. They asked if there was anything they could do for us, and Troop yells real loud—I couldn't believe it—yells out, 'Just give us your daughters!' Man, I wanted to get out of there quick. They were still friendly, though. They thought their place was gonna burn just like their neighbor's did."

June 18

After clarifying plans with the Cat skinner we went to work causing all kinds of destruction. I see Cat lines in general, and in Alaska particularly, with its fragile permafrost, as a last resort. At the time, though, with homes in the area and our backs to the wall, I decided to go ahead in hopes of stopping the fire at two thousand acres.

We tried to follow the actual perimeter, but still left areas of unburned green between the line and the fire itself. It was a clean line, twelve to twenty feet wide, cleared down to mineral soil. The debris was shoved to the outside, leaving as little fuel as possible for the fire when it hit the line.

Al darted back and forth in front of the Cat, waving at the operator with his hard hat. Roaring and crashing, the Cat charged over the spots where he had just been. It was unsafe as hell, but we both knew that. Just when I was certain that he'd been buried under a pile of slash, Al's head would pop up quick as a weasel, and he'd be waving his hat again.

After two hours, we tied into the saw line. I walked the Cat back out through the burn to the Cat road, clearing an escape route.

"Thank you, sir," I said, shaking the skinner's hand. "Damn good job! And we didn't kill anybody, either. As far as I'm concerned, you're free to go, but I'd like you to hang around out on the road. Fuel this old girl up. We'll probably need you later."

"Let me know if you do," he said. "I know these folks back in here. It ain't pretty to see your neighbors burned out." Without another word he climbed back on the Cat, flopped down in the seat, hit the throttle, pulled the right friction, and spun the huge tractor around on a locked track. When he released the friction, the big machine lurched away.

Al had gone ahead to join Boatner and the others. I finished the Snickers bar and choked down some jerky that chewed like salty rocks. My knees ached worse than ever, especially my right one. I was soaked in sweat, but I felt invigorated and very much alive. I looked at my watch—6:20.

I took off in the direction of the jumpers and found them sleeping, curled like a bunch of pups. I lay down next to them. The ground was soft and cool and felt wonderful. I lay there a few minutes, my eyes closed against the gathering daylight.

"Taylor, Romero," blared my radio. I bolted upright. I wrestled from underneath my jacket, groping for it.

"Taylor, Romero," the radio demanded again. Some of the jumpers stirred, mumbled profanity, then rolled over. I scrambled to my feet, turned down the volume, and walked a distance away from the sleepers.

"Romero, this is Taylor. Go ahead."

"Yeah. How's it going over there?"

"Not bad. We got something to make a stand on. Some road, some river, some saw line, a mile of Cat line."

"Sounds good, Murry. Thanks for the good work."

"I sent the Cat to the road to stand by. The guys are down. We'll be up in an hour or so. We'll spread out from the jump spot to the head. Did you get a weather forecast?"

"They said we should have one by ten o'clock. They're setting up a command post at Clear. The fire's gone Class II."

I filled Rene in on what I had seen at the head, then went back to the jumpers, turned off my radio, and lay down again. After a miserable half hour, I got up and walked over to the fire line. The sun was well up by then. I made a little campfire and sat down. It wasn't long until a long shadow stretched up from behind me.

"Hey, Boats," I said.

"Morning," he said. Boats was red-eyed, stiff-legged, and smelled of sawdust.

"Good job on the Cat line."

"The boys pretty beat?"

"Trashed is more like it! We worked straight through until the Cat went out to the road. They been down, oh, about forty-five minutes," Boats replied.

I filled my canteen cup with water and set it at the edge of the little fire. Boats banged his cup against his knee to clear the sawdust, poured in some water, and set it next to mine.

"So what do you hear? I couldn't catch much on my radio with the saws going."

"Well, hopefully, Rene has enough fire trucks to hold the tail. There's more on the way. Be here by noon maybe. We have this Cat," I said, motioning toward the road with my head, "and supposedly another due anytime . . . although I'll believe it when I see it."

Boats just sat there staring into the fire as if he could see through it.

"From what I gather," I went on, "Troop's contained his flank for the time being with saw line and some hose lays near the tail. But the saw line is skinny and crooked like ours. He'll be up to his ass in alligators if he gets an east wind, and we'll be in the same shape if it comes from the west."

Boats shook his head in vague agreement. He knew what we were in for.

"Two coffees or one?" he asked.

"Two." Stirring our coffee with sticks, we sat quietly for a few minutes.

"Tell me," he finally said. "Why are smokejumpers such idiots?"

I blew the ashes off the surface of my coffee, took a sip, and instantly burned my tongue.

"Idiots!" Boats repeated. "Why do you think that is?"

"Are you talking in general, or do you want to get specific?"

Boats swiped his nose across his sleeve. "What is it that keeps us in this miserable job?"

I sat thinking and chuckling to myself.

"Most *normal* people," Boats went on, "are at home sleeping in beds, with pillows and sheets, and maybe even other *people,* we're out here acting like a bunch of brush apes, running chain saws all night, swilling coffee that tastes like battery acid, eating stale candy bars, spilling gas and oil all over ourselves, filling our eyes with sawdust, ruining our hearing, and then lying in the dirt like a bunch of pigs, snoring and farting. I'm saying it's not normal. This shit is *not* normal!"

I sat staring into my coffee and wondering about the people who'd lost their beds when their house burned.

"Pretty morning," I said finally, sweeping my hand toward the Alaska Range and Mount McKinley. Boats turned and looked. His face softened.

"Yeah," he said quietly. "You got me, there."

Tom Boatner grew up as the son of General Boatner, U.S. Army. Coming of age during an era of war protest, Boats left home and journeyed north to Alaska to fight forest fires on the Tanacross helicopter crew. He and his fellows took sacred vows to outperform the smokejumpers, who, in their view, were little more than a bunch of pampered, overrated prima donnas. Three years later, after working with jumpers on several fires, Boats changed his mind and decided to become one. That was ten years ago. By the time of the Clear fire he was our crew supervisor.

"When's your trip?" I asked, motioning toward the mountains again.

"Ten days. Two weeks from now we'll be seeing a lot of white."

Much of his life, Boats had dreamed of climbing Mount McKinley, the tallest peak in North America.

"Excited?"

"I've wanted it a long time. And now it's finally coming together."

I thought to ask if he was expecting to find a ready supply of "normal" people, houses, and warm beds on Mount McKinley. "I don't know, Boats," I said. "People do funny things."

A few of the other jumpers walked over to our fire, looking like peevish cavemen.

"Morning, boys," Boats said, galvanizing his composure. "How's everybody doin'?"

"Coffee," Dewitz grunted, reaching down to scratch his crotch. "I need coffee."

We made two more campfires.

While we heated cans of pork and beans, hash, and kernel corn, I told them what lay ahead and what I'd heard from Troop and Rene. Cans of fruit cocktail, peaches, and pear halves cracked open. We drank coffee, ate, and began talking and chuckling easily.

"Shit," Quacks protested. "There's weevils in my oatmeal. Live ones."

"Lucky you," Kubichek replied. "Carbohydrates turned protein. I'll trade you."

"It's a deal," said Quacks. "One weevils and apple for one raisins and spice."

Life and the new day caught hold and soon we were laughing and picking on each other as usual. Once we'd finished eating, I sent four guys down the road to check the burnout at the tail, the saw line around the slopover, and to tie in with the state fire trucks. Then I got the rest together for a briefing.

"We shouldn't have much trouble until the wind comes up," I told them. "When it does, keep your eyes open. Spread out, be ready. If it jumps the road or our line we'll have to be quick. Have some piss pumps standing by." Piss pumps, as firefighters call them, are five-gallon rubber bladders equipped with a spray nozzle and worn like a pack. A small amount of water at the right time in the right place can do wonders if you know how to use it.

"Let's be thinking safety today, boys," I said. "We may have a tough one ahead of us, so watch out for each other."

Taking our stations along the saw line and on out the Cat line to the birch woods at the head, we spread out, each man patrolling a four-hundred- to five-hundred-foot section. Such spacing enabled us to cover the entire right flank, yet still to be close enough to yell for help.

At suspected problem points adjacent to the unburned islands we'd left making the Cat line, we posted people with radios. With a mile and a half of crooked line, we had to patrol back and forth constantly in order to maintain visual contact. Once I had the crew lined out, I walked to the head again to extend my reconnaissance toward Troop. With the day warming, it was critical that someone scout the remainder of the head without further delay.

Rene called. The helicopter that was to have begun bringing in crews had broken down and sat grounded in Anderson. Another had been ordered. There was no estimated time for its arrival.

Perfect, I thought. Enter Big Ernie, the smokejumper god. Broken helicopters are one of his favorite tricks. Unfortunately for us, Big Ernie's powers are great and not subject to appeal. He'd no doubt consider praying as some kind of whining and line up a few days of record temperatures, high winds, and bad radios in return.

Out on the head, where the trees were still alive, waxy birch leaves flashed in the morning light. Sometime around midnight the fire had run into them. In thirty yards, it had gone from a raging crown fire to a creeping ground fire. Never before had I seen the fire-retarding effect of birch woods demonstrated so dramatically.

Black spruce, white spruce, and similar species contain high concentrations of flammable resins, which enables them to withstand the subfreezing arctic winters. Birch also contains resins, but not highly combustible ones. Also, the leaf litter, which makes up the major component of the birch duff layer, holds moisture well and quickly decomposes into soil only a few inches below the surface. Since the birch will not carry fire and the understory combusts quickly, fire typically slows down in birch. The night air, with its cooler temperatures and higher humidities, the proximity of the river, and the low flammability of the birch had combined to stop a wall of fire in a matter of minutes.

Once, in thick spruce fifty miles south of Galena, I watched a cow moose and her two spindly-legged newborns trying to outrun a fast-moving fire. A band of flames flanked them on both sides. Soon they disappeared in the smoke. She'd been brave, but no doubt had finally been forced to abandon her calves and flee.

To our great delight, however, later that morning after stopping the fire, we saw the cow and her two offspring holing up in a small island of birch. The fire was still smoldering here and there, but the three were in no danger. Did our intrepid moose understand basic Alaska fire behavior, or did she just get lucky?

Personally, I'll go with instinct, genetic memory, and the divine wisdom of motherhood.

A half mile beyond the end of our Cat line, I ran into some red flagging tied around a scorched birch. There was a message written on the flagging:

THIS IS AS FAR AS I GOT. 6-18-91 3:15 A.M.—TYLER ROBINSON—AK SMOKEJUMPERS.

I got on the radio.

"Troop, Taylor."

A light breeze rustled high in the birch. Small wisps of ash swirled out in the burn. Then dead calm. I was about to try Troop again.

"Taylor, Troop."

"Mornin', Troop. How's life on the left flank?"

"Hey there, Leathersack," Troop said calmly. "Let me tell you. It's just great over here."

"See any dragons lately?"

"Oh, man." Troop gloated. "We got her. I'm telling ya, her eyes are crossed. She's on her back. Her legs are sticking straight up. She's lookin' pr-r-e-e-e-ty good!"

"I found Tyler's flagging. The head's no problem, but I was wondering about your side."

"These guys are animals, I tell ya. Don't worry, we can hold her."

"Well," I said. "I've got some strange-looking creatures over here, too, but I'm worried about this side. Thick fuels, dog-haired spruce. With a wind . . ."

"Good deal." Troop cut in. "Sounds great. Let us know if you need a hand."

"Thanks, Trooper. Sounds like a fine job over your way."

"Yeah. These guys were something."

"Talk to you later. Taylor, clear."

"Animals, clear."

That was it. We had contained the fire. Now, we had to dig in and hold it. Rene had the tail and was traveling back and forth in a pickup assigning work to the state engines and keeping us posted as resources arrived. Troop and his animals had the left flank. The birch and the river held the head. My group had the rest.

Back on the Cat line, I noticed a few small dust devils dancing around in the burn. Cumulus clouds were building along the slopes of Denali National Park, twenty-five miles south. A light but steady breeze began kicking up occasional flare-ups. Along the Cat line, several pockets of unburned green were putting up a lot of heavy white smoke. I walked out to the road and met with Rene.

At eleven o'clock we radioed Fairbanks and put two air tankers and an Air Attack aircraft on alert. I talked to the Cat skinner. He was fueled up and ready. The engines on the road had watered down a good part of our burnout, but there weren't enough of them to cover the road, protect homes, and move up the right flank as well. It would be up to us to hold our section; the same went for Troop's side.

Rene and I discussed what we might do if the fire blew out, then we parted. I hurried back to see how my right-flank guys were doing. I hadn't heard them on the radio, even though the fire was heating up. When I got to them, they were doggedly battling one flare-up after another. Patrolling back and forth, they were knocking down hot spots with spruce boughs. Moment by moment the fire was making threatening runs at the line. Clouds began building overhead. When clouds form, they draw air into them from all sides, creating local winds. One was coming our way, pulling an increasing wind across our line and into the thick black spruce that surrounded the homesteads, the Air Force Communication Facility, and the small settlement of Clear.

At noon we had our first spot fire over the line. I called Troop to ask how things were.

"Squirrelly, but the wind's in our favor. I think we can handle it, but we're gonna need everybody we got."

I called Rene and ordered Air Attack and one air tanker.

By one o'clock the air was full of blowing dust and smoke. Hundreds of acres of the interior began rekindling. Our flank was the worst and was now the front line of the fire effort. Gusty winds blew thousands of sparks across the line and into the green. Heavy smoke made it impossible to see more than a hundred feet. I met Boatner, Seiler, Dewitz, and Wasser on the Cat line. They were soaked in sweat and gasping for air; snot ran from their noses, tears from their eyes.

"Air Attack's on the way," I yelled. "When it gets here, if I'm not in a position to direct it, you guys talk to him."

They agreed, shaking their heads, saying nothing, constantly keeping an eye out for more spots. The smoke was thinner and easier to breathe close to the ground, so we all crouched down.

"I'm going back up the line again to check the head. I saw it earlier, but this wind changes everything."

The jumpers were full of fight, but I had to wonder how long they could hang on, having been awake twenty-eight hours with possibly another twelve to go.

"This could get to be a real bitch," Boatner said.

"Get to be?" Wally squalled.

"The thing I'm concerned about," I said, "is how far you guys are spread apart. It's right on the edge of dangerous. This is what I want. Bunch up, so each man is in a group with a radio. Patrol. If things go hen shit . . ." Suddenly a large flare-up exploded with a roar not far away. We stopped and watched, wondering if this would be the one.

"At least," I went on, "then we'll be able to move out together." The entire group was coughing, and it was hard to keep their attention.

"Look," I said sternly, "Let's not get too much hero shit in our heads. Remember where the saw line meets the Cat line?"

"Yes!" they said.

"That's your escape route into survivable burn. It's a line I had the Cat put in this morning. Orange flags are tied where it takes off. It's not real clear at first, but follow the Cat tracks. If it gets too hot in here, move out. We can always catch this thing somewhere else."

Several flare-ups hissed and woofed in the smoke around us. Connecting with his eyes, I made sure Boats knew I was talking to the group in general but to him in particular.

"When the time comes to pull out, you're to be the one to get it done," I yelled. "Make sure everybody knows where the escape route is. Pull out in plenty of time so we know we got everybody. Remember: To this mindless fucking fire we're nothing but another source of fuel."

"OK, boss," Dewitz sang, mocking seriousness. "Hey, boss," he went on, hard hat turned sideways, cross-eyed, and staggering on wobbly knees. "What does it mean, boss? All this smoke, man. Like, what's it all mean?"

"No, he's right." Wally offered. "If we get fried, think of all the overtime we'll miss."

"Not to mention hazard pay," I said, walking away, leaving them to tear after another flare-up.

Down the Cat line I saw more and more fire. I found the next group working a hot little spot fire, eating smoke, coughing, and laughing at their predicament. I mentioned the escape route, then hurried on toward the head. When I got to where the Cat line met the birch, someone burst on the radio, gasping.

"Taylor, Kubichek." Electricity shot through me.

"Taylor—Go!"

"Murry, it's jumped the road bad. We need help, quick."

"Any engines there?"

"A couple coming, I think."

"Hang on. Heading your way!"

I started jogging in the direction of the road, cut through the burn, intersected the Cat road, and broke into a steady run.

"Taylor, Boatner."

Frantic, I stopped and jerked my radio from its holster.

"Go, Tom."

"Need help?"

"Hang on. Let me see what we got."

"Copy."

I jammed my radio back into its holster and ran through thick smoke. At one point, a blast of superheated air whipped unexpectedly across the Cat road, hitting me broadside with searing heat. I ran on, stooped over, hightailing it down the road. Then up ahead I saw it. An ominous black cloud rolling out through an area of open woods the size of a football field, pushed by thirty-foot flames. Dust and smoke swirled crazily in the Cat road as if to serve warning that the fire was on the verge of doing something bizarre. I don't know what it is that happens to firefighters at such times, but I remember being possessed by one singular thought—I would stop that spot fire if it was the last thing I ever did. All our work was at stake, not to mention the homes and the safety of my jumpers on the right flank. If the spot fire escaped, it could cut behind them and possibly draft the main fire into a blowup, catching my jumpers in the middle.

"Taylor, Romero. What have you got down there?"

"Spot fire! We need engines, quick."

"They should be there any minute."

"Copy. How far out is the retardant?"

"Hold on, I'll check."

Another radio transmission cut the air.

"Taylor, this is *363*. We're six minutes out with *Tanker 33* on our tail."

It was Pat Shearer in *Air Attack 363*. Pat knew Alaska fire and was a top Air Attack boss.

"Patrick, you got here just in time. Radio Fairbanks. Roll another tanker."

Four bodies materialized out of the smoke coming my way. Heads down, handkerchiefs tied over their faces, it was the four rookies and snookies I'd sent back to check the burnout—Kubichek, Quacks, Togie, and Persons. By the looks on their faces I could tell that they'd rather die twice than have to face me under the circumstances.

"Damn it to hell," I shouted. "You need a few good men, and what do you get? Rooks and snooks."

"It kept getting worse and worse," Kubichek shouted. "Sparks across the line, constant spot fires, then one took off before we could get to it."

"Shit, you've got to be quick!" I told them. "This is a forest fire, not a sock hop."

The fire and wind suddenly began buffeting the air all around us, making a terrible pulsing noise that shook the ground. In heavy smoke, the sun dimmed to a dull red. I considered the risk of entrapment. Much of the interior area of the burn was reburning hotly. What I'd once considered safety zones had now become enveloped in blowing fire and smoke.

Just then Air Attack cut in on my radio.

"OK, Murry, we're coming up on your fire. Any other aircraft in the area?"

"Not that I know of. Check with Rene. Yeah, give us a look, Pat, then we'll talk."

Two wildland engines pulled up—the ones that had been working the tail of the fire. I jumped on the running board of the first and explained what I wanted.

"You guys will be key in stopping this spot fire," I explained. "The air tankers will drop retardant along its outside edge. These jumpers and I will move in behind it. Follow up with a hose lay. If we don't get in there quick, it'll rekindle."

By the time I had them lined out, Shearer was back on the radio.

"OK, Murry, we've checked it out. Are you there at the road where it crossed?"

"Roger!"

"OK. That's your first priority. We'll try to angle it off to this other road that runs east. If you get people in there right away, you might be able to hold it."

"That's what I wanted to hear, Pat. Go to it."

"Looks like your next area of concern is just north, ahhhh, up on the Cat line."

"That's it," I said. "When you get to that, contact Boatner. They're in a lot of heat. Keep an eye out for spots behind them. They can't see much."

"OK! Copy that. We're going to get some altitude now, and I'll get with this tanker and go to work. How on that?"

"OK, Pat, the drop area's clear. We're with you."

Air Attack planes are sometimes called "lead planes" since they lead the big tankers in. They themselves drop no retardant. Piloted either by private contract or government pilots, they carry a specialist on board to coordinate the tactical effort.

Air Attack 363 cut up through a chrome yellow sky in a graceful arc, the sun flashing off its wings just before it disappeared behind a cloud of smoke. When *363* came out the other side, *Tanker 33* was right on its tail. Moments later, the air war began. In a do-or-die effort, the lead plane and the big DC-7 roared by two hundred feet overhead. Three accurately placed drops fell like pink rain, extending a retardant line a thousand feet along the south flank of the spot fire, temporarily cornering it between the north- and east-running roads.

Minutes later, more engines arrived. As soon as the retardant hit the ground, the rookies, snookies, and I led the charge in, knocking down flare-ups with spruce boughs. The engine crews were close behind, stringing hose from the road, adding lengths as they came, spraying and cooling the edge further. The four jumpers and I leapfrogged around one another, beating out hot spots and gaining ground as fast as we could walk—tears streaming from our eyes, snot slinging from our noses, coughing, spitting—taking advantage of the fresh retardant.

Half an hour later we emerged from the woods onto the east road, covered with a slimy, red substance that burned our skin and stung our eyes. The rooks, snooks, and I stood on the east road taking a breather and watching the fire trucks secure the spot fire.

"Damn fine job," I told them. High fives were passed around.

"Ah-uhhh," we proclaimed, celebrating the brotherhood of smoke-jumpers.

Kubichek took out a candy bar, tore it into five pieces, and passed it around.

"Don't feel bad," I said. "I know it wasn't your fault."

"Well," Kubichek said. "Everything was going so well. Just when it looked like we might have it, the wind came across the road."

In the meantime I'd heard nothing from the right-flank jumpers. From where we were, I could see their area engulfed in heavy dark smoke, there had to be spot fires over their line. Alarmed that the right flank might be on the verge of a blowup, I got on the radio.

"Boatner, Taylor."

"This is Boatner," he said, coughing.

"I don't know how well you can see, but things are getting radical over your entire area."

"Copy that," he said, still coughing. "We're holed up where the escape route joins the Cat line. For now we're fine, but . . ." Boats's radio cut out and then came back on, and I could hear shouting in the background. Then, "We may pull out, but it'd be worth . . . can . . . hold on . . . load of mud . . . move back in."

More shouting came from the background.

"OK, it's your call. Don't compromise your way out. There's a hundred-yard strip of green along your escape route that could go any time. See what I'm saying?"

"Copy. We got our eye on it."

Air Attack 363 zipped around, banking high on its wings, flashing orange and silver through the smoke. The second air tanker circled the fire, diving time after time headlong into heavy smoke, its large engines roaring in defiance.

A huge ball of fire erupted above Boats's area. I tried him on the radio. No answer. I called Pat Shearer in the Air Attack ship.

"I can't really say that I see anyone in there," Pat said. "The smoke's too thick. We'll swing around to the other side and get in low, maybe we can see better from there."

Just then the Cat came banging up the road. Through swirls of dust and blowing smoke, I ran to it, climbed aboard, and pointed north as we turned off down the saw line. Minutes later we pulled alongside a wall of fire forty feet tall and we stopped.

"Can we punch through to the end of this line?" I yelled. "The jumpers are in there."

The skinner hit the throttle, dropped the blade, and we began rolling the hot edge of the fire back on itself. Embers and sparks blew sideways in strong gusts as I held up an old canvas against the heat. Something caught under the Cat, pitched it forward, and stopped it dead. The skinner grabbed the clutch, shifted to reverse and shoved the clutch back in. The Cat rocked back and forth for a couple seconds, then clamored backward off two big spruce bent under the carriage. We glanced at each other. If we'd high-centered the Cat, it would have burned right where it sat, and we would have had to run out into the green and angle back to the road before the fire caught us.

Again, the skinner slammed the throttle, dropped the blade, and tore off into the smoke, me holding up the canvas, he with a cigarette clinched in his teeth.

Through the smoke I caught a flash of *Air Attack 363* up front about three hundred yards. A moment later an air tanker roared past, loud enough to be heard over the banging and clanging of the Cat, and a pink mist filtered down through heavy smoke. My radio was squawking about something, but I was too busy hanging on to the Cat to trouble with trying to hear it.

A quarter mile later we came upon Boats, Seiler, Dewitz, Wally, and the rest of the first group. Their eyes, runny and red, peered out from under the brims of retardant-stained hard hats. I crawled off the Cat and instructed the skinner to continue ahead, widening the line.

"We kicked its ass, boss," Dewitz declared. *"Kicked its ass."*

"Looks like it kicked a little of yours, too," I said.

"We were too busy arguing whether or not to pull out to notice the fire much," Boats said, looking accusingly at Dewitz.

"It was pure sweetness," Dewitz assured me. "A piece of fucking cake."

Boats looked at me.

"Without the Cat line and the retardant, we'd have lost this whole flank. The smoke was bad. What worried me most—we couldn't see. I don't know which was more trouble, the fire or Dewitz. I was thinking pull out, but no! Not Scotty. He wanted to argue about it. We hung in here knowing the retardant was coming, but don't worry, we were ready to move."

"It's still touch and go," I said. "Keep the same patrol strategy. Pump these engine guys up if you have to. Some of them are pretty green. They'll be bringing a hose lay in. Show them what needs done."

With another two loads of retardant, the arrival of three more engines, and some head-down, ass-up fire fighting, by four o'clock that afternoon the rampaging right flank was once again contained. About 3:30 Troop asked for part of the last load of retardant to reinforce a slopover on his flank. Within the hour his animals had finished cutting two helispots in the birch woods between the head of the fire and the river. The replacement helicopter began shuttling in crews from a staging area in Clear. Activity quieted over the fire, and as the sun swung down out of the smoke toward the horizon, the heat of the day passed, and we spread out along the line to watch for spots and grab a bite to eat.

By seven o'clock the road was alive with new arrivals: trucks, pickups, fire engines, another Cat, two nurse tankers. New firefighters, wearing clean, brand-new fire shirts, were laughing and having fun. Exhausted, the jumpers continued to work throughout the remainder of the afternoon and into early evening.

Just after nine o'clock, Rene radioed the line to relieve the jumpers from duty.

We walked off the fire that night feeling like kings, plodding the long two miles back to the jump spot. Forty of us had held a two-thousand-acre fire with just one Cat, two state engines, four EFF people, and some key air tanker support. The last stragglers dragged themselves into camp about eleven o'clock, just as an unexpected alpenglow flushed a devastated landscape into a rosy brightness.

It took a few minutes to collect our gear, set up our tents, and fall into them. No one spoke a word, just sudden collapse.

The next morning I woke to the sound of a Bell 212 helicopter as it flew by, ferrying one of the crews out to the line. I looked at my watch—6:30. Stiff and flat, my body felt like the D-8 had parked on it all night. I was beat-up, sore, dry-mouthed, and ragged.

Before long I heard the breaking of sticks and the crackle of camp-fires. The thought of a hot cup of coffee and getting back with the others brought me out of my sleeping bag, groaning and battered, but strangely refreshed.

Outside my tent I stood on a small rise and gazed out over camp. The fire lay in the distance, and beyond that the forest stretched south all the way to the Alaska Range. Except for a few sunlit openings along the skirt of the mountains, the sky was gray and overcast. I grabbed my PG bag and dropped down the little slope into camp. Already half of the crew was up.

"Good morning, you brush apes," I said.

"*Ah-uhhh,*" they agreed.

I poured water from a five-gallon plastic water container into my hard hat and washed my face for the first time in two days, then headed for the nearest campfire.

"Coffee coming right up," Troop said.

Boats stood next to Troop. He looked rested.

"This has been one hell of a great fire," he said. "Not bad for a bunch of idiots."

Dewitz came stumbling out of the woods. He looked like he'd spent his entire life rolling in an ash pit. Scotty's hair, what was left of it, stood straight out like the wings of a toy glider.

"A fine good morning to you, gentlemen. Look good! Feel good!"

"Feel good?" Mitch grumbled, stirring his coffee with a dirty spoon. "I feel like shit."

"One time I met this bum outside a bus depot in Medford, Oregon," Dewitz said, turning to Mitch. "You know what that old man told me? I'm going to tell you guys—especially you, Grandma. The old man said, 'Son, in life you can be happy or you can be crappy.'"

"OK, Mother Teresa," Mitch snorted. "Look happy, be happy, that's you. But me, I'll stick to crappy and crappy in both cases. How's that?"

"What about all the fun we're having?"

"You call this fun?" Mitch scoffed, rolling his eyes to indicate the rough camp and our general condition. "You need your head examined there, Scotty boy—maybe get your hair done while you're at it."

"I call it good fire fighting, is what I call it," Troop said.

"Ah-uhhh!" we grunted, in a chorus of male approbation.

10

On June 23 eight of us were released off the Clear fire and driven back to Fairbanks in a school bus. Arriving at the shack at five in the afternoon, I checked the jump list and called Sally. Nearly a month had passed since we'd met. I wasn't even sure she was still around. The phone rang several times. Sally answered. Awkwardly, I explained that I had wanted to see her again but had been out in the bush. She told me she had taken a job waitressing at the Captain Bartlett Inn. I asked her if we could have dinner.

"This isn't another one of your tricks, is it?" she asked. "You ran off last time. Remember?"

"No more tricks," I said. "I promise."

Ten minutes later I was in my barracks under a hot shower, shampooing my hair and watching globs of gray fall at my feet. By the end of the third shampoo, the suds stayed white. The hot water running over my skin was wet, warm, and sensual. Such pleasure from simple things—clean Levi's, a clean shirt, combed hair, a temporary respite from danger.

At the Chena Pump House, Sally and I took a table out on the deck by the river and sat in silence for a moment, taken in by the tranquillity of slow-moving water. Sally ordered us a large platter of steamed clams and two beers. I was barely able to take my eyes off her.

"What is it that you keep looking at?" she asked.

In her sleeveless yellow-and-white print dress, Sally was the girl I'd once seen in a painting. The image was of a girl in a straw bonnet—its brim ringed with flowers—dangling her bare feet off a bridge while fishing on a warm summer day.

"I'm looking at you and wondering what goes on behind that smile of yours."

"Now that's a new one, isn't it?"

"I doubt it," I said. "But if it is, it shouldn't be. Not everyone has a smile like that, you know."

"Well, thank you," Sally said, smiling bigger than ever. "That's very kind. If I ever happen to figure out where smiles come from, I'll let you know."

When the clams came, Sally devoured them, laughed, winked, and ordered us another beer. We agreed that it would be nice to spend some time together. After a while I began to get sleepy. About midnight I drove Sally home and, after a long hug, headed straight for the barracks. For the first time in over a week I fell into a real bed, one with a soft mattress, clean white sheets, and a soft fluffy pillow.

The next day started with roll call, PT, and a weather report predicting heavy lightning over a large part of southwestern Alaska around Aniak and McGrath. Lightning was also forecasted from the headwaters of the Fortymile River to Lake Minchumina, and south of the Brooks Range from Bettles east to the Sheenjek River Valley. The jumpers seemed edgy and tense.

"Why don't they try saving some paper," Seiler grumbled, "and just say lightning over the whole state."

"Weather nerds," croaked Erik the Blak. "They get off playing games with their little maps and big words. All they really do is go in the back room and toss some chicken bones on the floor and then come back and tell us the weather. Modern-day witch doctors is all they are."

At noon I called Sally to confirm our plans to go dancing that night at the Howling Dog. After lunch a little voice in my head suggested I check the lightning monitor. Around the corner from the operations desk, a map of Alaska with a dot grid overlay appears on the computer screen. The grid points indicate the positions of BLM lightning detectors across the state. These remote sensing devices operate by solar power to pinpoint ground strikes, then relay the data via earth-orbiting satellites to a computer like ours. Cloud-to-cloud lightning is not recorded. Only ground strikes cause the little machine to throw its tiny fit and post its latest data.

The computer triangulates the input signal and then places a small x on the map locating the last strike. The longitude, latitude, and time of the most recent strike are promptly listed in the upper right-hand corner of the screen, as well as the total down strikes for the day. Using this information, the fire management zones plan detection flights to track the storms and locate new starts.

Already the total number of down strikes was at 650, and they were distributed roughly as predicted, the bulk being between Fort Yukon and Bettles. Every ten or fifteen seconds another small x would appear on the map. When the lightning machine goes over nine hundred, it's best to be listening for the siren. My chances of spending another evening with Sally were being blipped away, minute by minute, one small x at a time. Still, I was way down the list at number forty-two—maybe the storms would be wet and drench the fires they had started.

Between three and five o'clock, we rolled four loads. I called Sally and apologized. At eight another load rolled west to Galena. Everyone left was held on standby until eleven. We waited, certain we would be flying any minute, but at eleven we were put off for the night.

The next day before work I tried calling Sally. The phone rang a long time before I hung up. People paced this way and that in the standby shack, checking gear on the speed racks, crowding the operations desk, staring at the jump list, reading the teletype, asking questions. After roll call we were briefed on the fires of the day before. The forecast called for more lightning. There were a lot of fires—so many that our dispatchers were changing the procedure for prioritizing them. Only three loads were left in Alaska—one in Galena, two in Fairbanks. Fifty additional jumpers had been ordered from the Lower 48 and would be arriving in time to jump later that evening. The morning briefing ended with a staid message from Rodger Vorce, our boss.

"I'll make this to the point," he said, scanning the group. Rodger had on brand-new Levi's, a white short-sleeved cowboy shirt, and Wellington boots. He had just returned from the Puzzle Palace where, in a cramped back room, the lead dispatchers and fire managers had conferred.

"I guess you guys are seeing it. The country's drying out. May went down as the driest on record, and so far this is the second-driest June. We're already seeing peak July burning conditions. If your fire lies down at night, it probably won't stay down come morning. During the heat of the day, some of them are probably gonna scare the hell out of you. Be careful with your gear. Some of these fires are going to roll no matter what. We've already lost gear, and when we start losing gear that tells me something. I'm hearing stories I don't like. These burning conditions are the kind that get us in trouble. By tonight we'll have a lot of outside jumpers here, some for their first time. You just might wind up with a planeload of them jumping some gobbler."

Pausing for a moment, Rodger rubbed the back of his neck. "Take some time to help them understand our fire behavior and the way we do things. It'll pay off in the long run."

Putting his thumbs in his belt loops, Rodger rocked back on his heels, tongued his chew of Copenhagen, and looked at us sternly.

"This is what it comes down to. Last year we had too many close calls, and now we're already having them again. We're all here and we're all fine this morning. Let's keep it that way."

Rodger turned and disappeared into his office.

"Anybody got anything else?" the box boy asked. No one did.

"First load," he yelled. "Do your run this morning either out by the golf course or on the track at the units. We could get a call any minute."

Fire call B-335 came in at noon. A private plane had called in a small fire near the village of Fort Yukon. By the time we arrived, it had burned onto a thin strip of land in a narrow slough off the Porcupine River. It appeared to have burned itself out. We flew low over it several times, trying to spot a smoke. If we don't see smoke, we don't jump, unless it's in an area with large fire potential. After an hour of circling everyone on the plane was half sick.

Back in Fairbanks we put our jump gear on the speed racks, rinsed our faces, and piled into a van bound for the cookhouse. We'd just finished filling our trays and sat down to eat when another call came in. This one from the village of Venetie, sixty miles northwest of Fort Yukon.

From a jumper's perspective, fire B-361 had two problems: First, it already covered two hundred acres and was gaining steam—a difficult one to catch right away. Second, the Upper Yukon zone had taken a closer look at its maps and determined that the fire was not in a "modified" area, as first thought, but just across the boundary in a "limited" zone.

Wildfires in Alaska are managed according to established fire plans. The land is divided into four categories: critical, full, modified, and limited.

Critical fires involve areas where life and property are at risk: army and air force bases and other national defense structures, such as radar and communication facilities; as well as Alaskan native villages, small towns, and the outskirts of major cities. Critical fires are always aggressively suppressed.

Similar to critical areas, fires in full zones require full suppression action. Mining and other remote camps or research facilities fall into this category as do some designated natural resource areas—particularly river corridors, where for various management reasons, fires are suppressed during certain times of the year.

Modified areas may be allowed to burn if considered appropriate by the managers. Nearly all modified areas convert to limited during the second week of July, after which it becomes unlikely that, with the onset of cooler nights and higher humidities, fires will spread to troublesome sizes.

Fires in limited areas are allowed to burn all year. Here again, however, managers can decide to stop them altogether or to manage their direction of spread. At times, jumpers are dropped to protect remote cabins and homesites in limited areas. Nearly two-thirds of the entire fire protection area of Alaska is classified as limited.

Traditionally, the wildfire establishment was steeped in a fire control mentality. Fire was bad. A trustworthy Smokey Bear appeared on poster after poster surrounded by burned trees to warn us of the evils of fire. By the mid-eighties, the environmental movement had begun arguing the wisdom of allowing fires to burn naturally. Both groups eventually agreed that total suppression was no longer compatible with nature's way. That part was easy. The hard part was determining how much fire should be allowed to burn where and when.

In the case of B-361, the Upper Yukon zone decided to let it burn. They also directed us to fly to Fort Yukon and stand by. We landed, spit on the ground, cussed fire management policies, refueled the airplane, and then force-fed our queasy stomachs. After dinner I visited with the good folks at the Fort Yukon station. It was 89 degrees at six in the afternoon, with several thunderstorms out to the northeast toward Canada. Beth Greycloud, the local dispatcher, showed me how the lightning monitor had totaled 2,006 strikes during the past eighteen hours—most of them since noon.

I went outside, lay down on the lawn, and stared up into a dark sky. I was nervous and edgy. I tried to relax but couldn't. Thunder rumbled across the country off to the north. Wind rustled in the cottonwoods; the siren blew.

Moments later, drenched in sweat, we were suited and aboard the jump ship, roaring down the runway. We crossed the Christian River and flew west up the Chandalar and over the headwaters of the south fork of the Koyukuk. Three fires had been reported in the foothills of the Brooks Range along the southern boundary of the Gates of the Arctic National Park.

At 8:30, Kubichek and I jumped fire B-378. It had been started by lightning a few hours earlier and had received a little rain. It was a small, straightforward fire, so the spotter let Kubi take charge. About ten o'clock another jump ship came by and Erik the Blak, the spotter, dropped us extra hose and a bigger pump. We set up a Mark III pump on the bank of a nearby creek and pumped twenty-five hundred feet uphill to the fire. That kept us busy until three the next morning. After that we took time for a campfire, a bite to eat, and a good look around. If the fire was ordinary, the country was not.

The sky in the south turned dark and spit lightning. Rainbows appeared, then faded away.

"Think I'll turn in," Kubichek said.

"Yeah, me too. Good night."

Minutes later in my tent, I burrowed deep into the silky comfort of the sleeping bag and lay there listening to the far-off thunder. Just as I was dropping off to sleep, the first tiny drops of an early morning shower began tapping lightly on my rain fly.

We got up at eight that morning. The sky had cleared. Raindrops clung to the trees, brush, and grass as glimmering pinpoints of colored light. Our world was one of absolute and glorious quiet, but I knew it couldn't last. There was too much work to do. We spent the rest of the morning pumping water and dragging hose back over the entire area, putting out the last smokes. It was a long, narrow fire, running up the hill away from the creek, so it took a lot more hose than a regular half-acre fire. While Kubi spent time manning the hose, I began retrieving cargo chutes, then cut a helispot in a small clearing near the creek. About one that afternoon, a detection flight came over and told Kubi that Fairbanks was eager to get all jumpers back to town. Kubi reluctantly agreed to leave the fire later that night.

We began to hustle, making sure the fire was completely out, breaking camp, packing up, and hauling all our gear to the helispot—approximately six hundred pounds. A seasoned jumper knows not to strike camp until you hear the helicopter coming, but in this case, the orders were clear so we went ahead and did it anyway. Barely a heartbeat after we'd finished packing everything to the helispot, the detection ship came back.

"OK, you guys, this is the scoop from Bettles. They wanted to get you out tonight, but they can't . . . helicopter unavailable. Will plan on your demobe tomorrow around noon. How on that?"

"Sure enough," Kubi said, grinning. "That's fine. We can stand another night out here. Just as long as you don't forget us."

Celebrating our good fortune in getting to spend another night in such a pretty place, Kubi and I went to camp early and sat around our campfire talking. Somehow the subject of marriage came up. Kubi seemed doubtful when it came to the concept but deeply curious, nonetheless.

"You were married, right?" he asked.

I told him that I was. Then we sat quietly awhile, I guess both of us thinking about what to say next.

"The way I see it, marriage is like a lot of other things. You pretty much get out of it what you put into it. Why? You thinking about getting married?"

"Hell no! Not anytime soon I hope—I'm only twenty-five. I've got some things to do before I think about settling down."

"That was the problem with me, all right. I got married when I was twenty-four, in August of 1965—the same year I rookied—much too young."

"Jumpin' dog shit," Kubi yelped. "I wasn't even born yet."

Things got quiet after that. Thunder came from somewhere back in the Brooks. Lying back and resting my head on my PG bag, I began thinking about the last years of my marriage.

It was the sixties, and my new wife and I had spent our first three years living happily in the mountains of Northern California. In 1969 our son, Eric, was born and the three of us became the picture of a wholesome young family. The excesses of the sixties initially turned me off, but by 1970, when the Forest Service had taken us to Southern California, I began to get curious. The people we met there were weird but nice. In such a heady social climate, Kathy became more liberal than ever but never changed her basic commitment to the ideals of marriage. I, the conservative country clod, began smoking pot and immediately went off the deep end, talking philosophy and indulging a potent capacity for self-serving rationalizations. I read books like *Games People Play, Open Marriage, The Art of Loving,* and *Apes and Husbands*. Kathy and I attended parties where people dressed like gypsies, took LSD, and had "dropped out and turned on." One couple claimed to have Timothy Leary hidden in their barn. The cops had been there, they told us, but had left empty-handed. It was all very intoxicating. The next thing I knew I was having an affair with a forty-year-old Hollywood woman who owned a dress shop in downtown L.A.

During that time I attended encounter-group sessions at the Center for the Studies of the Person in La Jolla. In November 1972, I quit the Forest Service, and Kathy, Eric, and I spent that first winter venturing down the west coast of Mexico as far as San Blas, living on the beach in a small trailer, enjoying weeks of sunshine, emerald green surf, brown bodies, and our son, a wild child playing in the sand.

In the spring we made our way through California visiting family and then migrated up the Oregon coast to Seattle where we purchased six hundred dollars' worth of food and packed it carefully into the pickup. We traveled up the Alcan, taking twelve days to reach Fairbanks and our new life in the north. Kathy got a job managing the Tanana Valley Campground, and I signed on with the Alaska Smokejumpers. In mid-July I was sent south to Oregon and Washington to jump fires for nearly two months. While I was there, my wife of eight years found out I was having an affair with a Nez Percé woman. She wrote me a letter asking me for a divorce.

We broke camp at nine the next morning and packed our gear back to the helispot. The day was too hot for mosquitoes, and so we lay in the sun with our shirts off.

"Fire B-378. This is helicopter *92 Romeo*."

"*92 Romeo*, this is 378."

"We should be coming up on your fire shortly. Can you hear us?"

"Negative."

"OK—we should be in earshot soon. Let us know when you pick us up."

Within the minute we heard the unmistakable distant shudder of a Bell 212 pounding its rotor blades, coming our way. In the right conditions a Bell 212 can be heard for thirty miles, whopping along, echoing off mountains, radiating ever closer, bringing with it the prospect of more adventure. Kubi and I scanned the empty sky as the pounding grew louder and louder until at last a small speck appeared above the horizon.

"*92 Romeo*—Fire 378."

"This is 92."

"OK, got you. We're at your eleven o'clock, about three miles."

"Copy that."

The helicopter landed and shut down. We couldn't fit in all our gear, so the pilot decided to pick up the rest when he came back for the other jumpers from our Fort Yukon load, who were on another fire ten miles north. We loaded the helicopter to the ceiling, filling every available space, then wedged ourselves into a position where we could somehow strap on seat belts. As the helicopter began its run-up, I reached over, slid the door shut, latched it firmly, gave the pilot a thumbs-up, and stuffed in my earplugs. The pounding of the main rotor shook the entire load roughly as it strained to clear the ground.

We flew west down the south fork of the Koyukuk, along the southern edge of the Brooks Range, passing south of Wild Lake and the Wild River country. Ahead I could see the canyon where the John River leaves the mountains after flowing from its headwaters near Anaktuvuk Pass, some 130 miles north.

On the ground at Bettles we off-loaded our gear. The helicopter shut down, refueled, and flew back east. Bettles has a five-thousand-foot gravel airstrip, the Bettles Lodge, a small private air service, and an FAA flight service station. Its eighty or so summer residents are some of the most northern souls on the continent. BLM operates a dispatch office, a cookhouse, an emergency fire warehouse, and a parachute cache during the summers. A small cluster of wall tents—built on wooden floors—house firefighters, pilots, and other BLM transients overnight. In front of the Bettles Lodge there's a sign:

WELCOME TO BETTLES FIELD, ALASKA.
35 MILES NORTH OF THE ARCTIC CIRCLE.
POPULATION, 51—MORE OR LESS. ELEVATION, 643 FEET.
LOWEST RECORDED TEMPERATURE, 1975, −70
HIGHEST RECORDED TEMPERATURE, JULY 1955, +92
HIGHEST RECORDED MONTHLY SNOWFALL,
OCTOBER 1945, 35 FEET, 4 INCHES.
AVERAGE MEAN TEMPERATURE, +21.

No sooner had we stacked our gear on the dock in front of the parachute cache than Kubi began talking about food. Just then *Jumpship 17* touched down at the west end of the field and taxied to where we were waiting. Our ride back to Fairbanks, we thought. As soon as it stopped, Bob Quillin hopped off the tailgate with a clipboard in one hand and a packet of maps in the other.

Bob weighs about 140 pounds and stands five feet seven. "Body of a chicken, heart of a lion," the jumpers like to jest. They enjoy reminding Bob of a billboard depicting a skeleton chicken as it points to a can and proudly proclaims, "That's my meat." Thus, Bob has been saddled with the nickname Stewing Chicken. Stewing Chicken or not,

Bob is always ready for a round of banter when dealing with the crew. On that sunny morning, Bob came stalking our way, feathers ruffled. His personal bent is to immediately condemn anything that might lure jumpers away from the *joyous* tasks of hard work and suffering. This he accomplishes by prefacing his verbal assaults with what sounds like a poorly made duck call.

"Waahhk," he said, smiling sweetly, "Guess what I found for you girls."

We looked up at him and groaned.

"Cutest little . . . well . . . *kind* of little . . . fire you've ever seen. Up on the south side of Caribou Mountain in some real pretty country."

"We're hungry," Kubi said.

"Forget hungry. You ate all winter, didn't you? We got fires to put out. What do you think this is, anyway, a fat farm for pink-skinned, over-indulged, spoiled brats of America?"

In 1976 our venerable Stewing Chicken came perilously close to death on the Seward Peninsula. During a lightning bust near the coast, eight jumpers dropped in high winds near a hot, fast-moving tundra fire in what later became known as "the Golovnin Bay Massacre."

Bob hit the ground blowing backward at fifteen miles per hour, tumbled end over end, and was knocked out. As he regained consciousness he found himself being dragged cross-country by a fully inflated parachute, one arm tangled in the suspension lines and the other paralyzed. As he was wrenched up and over the tundra tussocks, Bob's arm jammed repeatedly between his body and the tundra, twisting and breaking, along with some of his fingers.

From the jump spot, Bob was dragged two hundred yards onto the beach. Across the sand he sailed. Just twenty feet short of the water's edge, his chute snagged a clump of driftwood. If not for such good fortune, he would have been pulled into Norton Sound. Considering the speed and direction of the wind, the jumpers would have had no choice but to stand by and watch helplessly as Bob was swept to sea.

As it happened, they got him to Nome by helicopter, and a Learjet from Fairbanks rushed Bob to the hospital in Anchorage. Eventually, he mended his broken arm, regained the use of his hands, and returned, spunky as ever, for another fifteen years of jumping fires.

Now, standing sprightly before us, Bob was all business concerning the fire on Caribou Mountain.

"This fire's already got four jumpers on it—I just dropped them. Some Redding and McCall guys. You should have a good chance at it. We'll drop extra hose and fuel for the pump, but we can't swoon around here all day like a bunch of sissies whining about food."

The flight to Caribou Mountain took twenty minutes. Bob told me the fire boss was a Redding jumper named Foley and suggested I make the decision whether or not to take over. When we came in over the fire, a half-mile band of three- to five-foot flames was backing across a broad tundra flat, eating its way steadily toward the south slopes of Caribou Mountain. Stringers of dark green spruce fingered delicately up the mountain in between light-yellow ridges covered with caribou moss. The first priority would be to stop the fire before it reached the steeper slopes. The remaining perimeter had burned hot in forty acres of black spruce but ended up smoldering in wetlands at the edge of the Kanuti National Wildlife Refuge.

As we circled I spotted the four jumpers on the ground. They apparently had seen things the same way because they were spread out along the band of flames. They weren't having much luck, though. Too much fire. They'd beaten out portions of the fire's edge with spruce boughs, but the line remained smoky and kept flaring up behind them. I would jump first.

Bob came crawling back from the cockpit and immediately we launched into a typical spotter-jumper shouting match just inches from the open door. The Stewing Chicken got down on his hands and knees and drew with his finger on the steel-plated threshold of the door. He was trying to tell me something about the fire—something he apparently considered very important. My eyes watched his fingers crisscrossing the floor, glanced now and then to his face, then out the door to the fire, then back again to the blank floor. I had no idea what he was talking about. Finally, he shot me a big smile and I smiled too. Not because I understood, but because I liked him so much. Having exchanged not one shred of information, we stared out the door at the fire as the plane banked left for our first streamer pass.

We jumped at two in the afternoon on a sunny day with light winds. The ride down was beautiful, with Caribou Mountain to the north and Kanuti Flats south. I came in for a nice soft landing in the middle of a big patch of blueberries. Kubichek—my jump partner—along with the rest of the entire load, landed about a hundred yards to the east on the opposite side of the creek.

"Hey, Leathersack, you knucklehead," Seiler yelled. "What are you doing over there? The jump spot's over here!"

Ha, I thought to myself. *So that's what Quillin had been trying to tell me.*

We set up a Mark III pump at the creek and ran a hose lay toward the jumpers on the mountain, spraying out fire as we went. By late afternoon we had worked our way up the line to the other jumpers.

About seven o'clock, fire boss Foley and I met. He seemed completely comfortable with the situation, so I saw no reason for him not to continue running the fire. He recommended we break for a bite to eat and have a few people start retrieving parachutes and gear. The perimeter of the fire had been secured, and the advancing band of flames had been doused with four thousand feet of nylon fire hose. The remainder of the fire had corralled itself out in the wetlands.

We rigged up a long-range antenna, and Foley called Bettles. They wanted to replace all the jumpers as soon as possible. Hearing that, Foley ordered two EFF crews complete with support gear and food for three days. Foley would stay and run the fire. The rest of us would leave as soon as our replacements arrived.

We worked the line until two the next morning then returned to the jump spot, where we made small talk, ate some more, and went our separate ways to set up tents and nab what was left of a night's sleep. On a small bench of higher ground just above the creek, the caribou moss was level and the elevation provided a commanding view of the fire, Caribou Mountain, and the Kanuti Flats. Near its edge lay the skeleton of a caribou, complete with skull and antlers, all bleached white as chalk.

Lying back on their curl, the antlers had settled down into tundra surrounded by caribou moss and miniature bearberry. Tiny red and white blossoms poked out of the eye sockets while patches of orange and green lichen spread upward in the bold, graceful curves of the antlers.

I set up my tent nearby, shook out a cargo chute for additional padding, and set stones around the perimeter to hold everything in place. I moved reverently, taking my time and thinking how fortunate I'd been to come upon such a scene.

The sky to the north was a fine pastel pink, leaving Caribou Mountain a lone gray sentinel looming above us. A loon cried far off in the flats. Now and again my ears picked up the wing flutter of the Wilson's snipe—a soft, pulsing whistle that rises in pitch through each series of beats.

I crawled into my tent, undressed, and snuggled down inside my sleeping bag. A pair of Wilson's snipes swooped overhead, the air whistling around their tail feathers as they performed acrobatic displays for their mate. I thought about Sally and when I might see her again. Possibly never. Women like her don't go unnoticed very long in Alaska. Three nights up late had me worn out and feeling blue. I folded a corner of my sleeping bag and cupped my hand over it and thought about Sally's breasts and how they looked beneath the yellow-and-white print dress she'd worn at the Pump House.

I woke that morning heavy and stiff, feeling like I'd died, gotten rigor mortis, and then returned to the world of the living. As I crawled out of my tent, I thought again that maybe I was getting too old for jumping. The fires seemed awfully hard, but then I couldn't remember when they hadn't.

Caribou Mountain stood tall in full morning sunshine, while the Kanuti Flats lay to the south as flat and shiny as a mirror. Woodsmoke scented the air. Never do I leave a wilderness camp that I don't take a moment to witness a scene of transitory domesticity. Like all human beings, smokejumpers have simple needs—to eat, to drink, to do good work, and finally, to rest and seek comfort. Playing out the satisfaction of those needs, we make long excursions to wild places, where we spend the end of each day engaged in homemaking, however crude. Violent weather, isolation, extremes—all these raise the value of warmth and security.

Around the morning campfire, everyone was in good spirits except for a Redding jumper who had hurt his back on landing. He looked miserable. We lingered at the campfire, drank extra cups of coffee, and ate as much as we wanted, much to Kubichek's pleasure. Foley radioed Bettles while the rest of us waited in camp for the helicopter, watching a pair of Canada geese fly west, becoming small specks in a big sky.

11

Fortymile Mountains

On our approach into Bettles, I twisted around and looked forward through the cockpit of helicopter *92 Romeo*. On the ramp sat the same jump ship we'd met the day before. Déjà vu—*Jump 17* and another spotter with a clipboard. This time it was Jim "Oly" Olson.

"I think we got a fire call," he jabbered, "Better get your rigs ready, we should be hearing anytime, you wouldn't believe where I've been, there's fires all over the place, they got a hundred extra jumpers from down south and they're ordering more."

Oly's eyes were golf balls with tiny blue dots for pupils. His red bushy hair looked like he'd flown all the way from Fairbanks with his head out the window.

"I gotta get away from this damned airplane," he said unblinking. "But everywhere I go—everywhere—there's another bunch of jumpers and another fire to spot . . . Quillin tricked me, that fucker. Now I'm stuck spotting. I've got to get away from it so I can get on a fire and get some rest."

"Get on a fire and get some rest?" Kubi wailed. "Are you crazy? I haven't had fifteen hours' sleep in five days, and I'm about starved to death. Which way's the cookhouse?"

Oly grabbed his radio, stared for a moment, and then began to talk to it as if it were alive.

"Dispatch—this is *Jump 17*. We need some food. How on that?"

We finished rigging our harness with new chutes and loading *Jump 17*, piled in the van, and made a mad circuitous dash around the end of the runway heading for the cookhouse. The van slid to a stop in front of the cookhouse in a cloud of dust, gravel spraying out from under its wheels. The cooks looked on astonished as we ripped fried chicken off the bone and drank gallons of juice. We drank milk, ate cake, pie, ice cream, and mashed potatoes. Kubi belched loudly as he wrapped pieces of greasy chicken in napkins and stuffed them into his pockets.

Fifteen minutes later, against serious protest, Oly herded us back into the van. We drove back to the plane, crawled under it, and lay down on our jump gear. *Oh god,* I thought, *just give me an hour so I can catch a nap.* Five minutes later the call came in. At 2:50 we lifted off from Bettles and were given the details. Four hundred and fifty miles south there was a fire in the Fortymile Mountains, not far from the small mining settlement of Chicken. I would be fire boss. After stuffing in my earplugs, I pulled on my helmet, shut my eyes, and tried to settled down for some sleep amid the chaos of *Jump 17* screaming along at 170 knots.

Lying there, something began to bother me. How was it, that in the midst of so much beauty and excitement, something still seemed missing in my life. I missed my winter home. More and more I found myself missing the friends that I had scattered through three decades of coming and going.

My family split up following the fire season of 1973. We left Alaska that fall in a lot of pain. The slow trip back to California was made knowing that our lives together would end as soon as we arrived there. Kathy and Eric moved to Fresno where she found work. In a friend's backyard, my wife and son set up housekeeping in the same trailer that we, as a family, had launched our dream of freedom in.

I drifted south and wandered the beaches of Mexico for two winters trying to comprehend the drastic turn my life had taken. My summers I spent in Alaska. In November 1974, I spent a month thirty miles north of Barra de Navidad on a beautiful tropical crescent of sand called Tenacatita Beach. That's where I met Sarah, a tall, blond, and shy nineteen-year-old. She was traveling with a group of friends she'd worked with the previous summer at Glacier National Park. I visited their camp every night, feasted on fresh fish, and made friends with Lucky Black; his wife, Toaster; Carolyn; Flake; and Nate—all Sarah's pals from Duluth.

Sitting around the campfire I could hardly take my eyes off Sarah, her breasts thinly veiled beneath her cropped tank top, the waves pounding and washing in and out against the sand. We became friends—not lovers—just walkers on sunset beaches, tellers of stories, and singers of songs. Though I enjoyed my time around her very much—and I did spend a lot—it was utter torment to not make the move toward becoming lovers. I simply couldn't do it. The longer I procrastinated, the more absurd the discomfort became. Besides the cowardice on my part, there was more to it than that. Sarah was precious and I didn't want to frighten her. She was young and inexperienced, a former ugly duckling turned swan. Our last night at Tenacatita, after our usual evening of campfire, we walked down the beach to her tent, and we sat watching moonlight flush silver over the Pacific. At last, I reached over and kissed her on the cheek. Taking my hand in hers, she led me into the tent. We curled up and spent our first night together. Wave after wave pounded the beach, thundering and hissing, rushing in and out, and I was completely happy to at last have Sarah in my arms.

The next day Sarah and her friend Carolyn left to fly back to Minneapolis, and I gave them a ride to Guadalajara. I told Sarah about Alaska and the work available on the pipeline being built from Valdez to Prudhoe Bay. I told her that if she'd come, I'd help her get situated and find a job. We hugged, she left. I figured that would be the end of it.

Six months later, on a warm afternoon in Fairbanks, the phone rang in the old standby shack. Sarah and a friend named Burgie were downtown at the North Star Bakery. My good buddy Johnny Culbertson and I raced down and found them sitting beside backpacks nearly as big as they were, drinking coffee, and eating veggie burgers. We spent the entire afternoon in the parking lot of the bakery, in my pickup camper, drinking tequila and celebrating. The arrival of two healthy, happy, suntanned female adventurers in Alaska is no small thing, even today, let alone back then when the ratio of men to women was twelve to one. I offered them my pickup and camper for a few days, while Johnny and I doubled up in a barracks room. In three days, the girls had jobs waitressing for a hundred dollars a night in tips. Saturday night the four of us went to a dance at the old Howling Dog in Ester.

We drank and danced and carried on like we'd just discovered gold. I finally passed out flat on my back on a big round table. After a while I came to and started dancing again, then passed out again—same table. Later, when I came to in the front seat of my pickup, Sarah was having trouble getting the gearshift in reverse. By the time we found reverse, I'd kissed her several times. We drove to the Tanana Valley Campground and parked in a camping space. Johnny and Burgie had ridden home in the camper, and by the time we'd parked they had taken after each other in the bed. Naturally, Sarah and I spent the night in their tent.

The next day I bought a twelve-by-fourteen-foot wall tent, and the four of us set up housekeeping under a stand of tall spruce. Sarah and I moved into the wall tent, and John and Burgie took the other one. We strung up ropes, covered the area with tarps, and nailed scrap lumber into crude furniture. Five years later, Johnny and Burgie married and started a family. They now have four terrific kids, one of them nearly as old as their mother was when Johnny met her. Sarah and I left Fairbanks that fall and traveled Alaska in my camper, eventually making our way to Valdez, where we caught the ferry to Seattle.

Within the month I had purchased forty acres in Northern California. We worked together cutting firewood and building a little shack down by the creek. It would be home until the main log house was finished. We took long baths in front of the fireplace in an old metal tub, skied by moonlight, met our new hippy-type neighbors, and fell in love. We skidded house logs with our new team of draft horses. We planted fruit trees and a garden, built a barn and corrals, drank wine by candlelight, rolled in spring grass, dreamed, and played like kids.

Each summer we returned to Alaska where Sarah got on a helitack crew fighting fires. In the fall we'd migrate back to California to continue working the land. By the end of 1978, I'd decided to give up jumping because my son, Eric, wanted to come and live with us. Sarah returned north in '79, and we spent the summer apart. That fall, she came home pregnant. There was a lot of crying and struggling, but eventually it gave way to forgiveness. I, more than anyone, could understand. I loved her very much; she was too good not to love. A few weeks later, she went with a neighbor friend to Ashland, Oregon, for an abortion. It was a regrettable decision. I should have asked her to marry me and have the child. Instead, I procrastinated, took too much of Sarah for granted, and the tension built. Then I made another mistake. I had a vasectomy without first allowing her a respectful say in the matter. Even though I did it partly because Sarah was having health problems with other forms of birth control, my vasectomy ultimately carried the message that I didn't want to have children with her. Things went downhill after that, ending at a sorrowful impasse. On a bright spring morning in 1980, heartbroken and sick, I stood in the yard, tears streaming down my face, and watched her drive away.

Aboard *Jump 17*, Bell and Seiler were peering out a window. Kubichek sat on a fire pack next to me, chewing on a chicken bone. The four closest to the cockpit were lying on the cargo with their heads back, eyes shut.

My left knee was beginning to throb. If I didn't keep it moving it hurt constantly. The pain seemed centered deep inside, a dull ache made keen by the vibrations that came through the floor. As I lay there struggling to get up, Kubi stuck out a greasy hand and pulled me to my feet. I flopped down on the fire pack beside him.

"It's just Fairbanks," he said, giving me a no-big-deal shrug. Fairbanks passed below us under a blanket of gray haze, close but far away, another world. Searching the city, I found the Steese Highway, followed it to Four Corners, the store, and the narrow dirt road that led to Sally's brother's cabin. Down through the trees, I could barely make out the roof of the cabin and the garage.

Fairbanks slipped from view, leaving me a small speck in a big sky, moving east across a great wilderness. I turned from the window, put my helmet back on, shut my eyes, and gave myself a little speech.

"Fuck it," I muttered. "You have responsibilities. Nothing matters except the job. Keep it together. Stay positive. Stay focused. This is what you train for, this is what they pay you for. It could last three months. Forget Sally. Thinking about her only makes it worse."

Just then two golf balls glued to a red mop dangled around the corner of the cockpit—Oly waving a piece of paper. The note came back.

> Fire B-449—Fire Boss Taylor—Fortymile—10 acres, burning
> hot—ridgetop. Ten miles N of Chicken. Work on Silver FM.
> Jump 19 en route out of Fairbanks.
> Air Tankers—124—Tanker 84—Tanker 33 rolling (also
> Fairbanks)
> Air Attack—17 Delta Lima. If 19 gets there first, the Blak will run
> the fire. If we do, it'll be you. Go ahead and do your checks.

What we saw in the mountains of the Fortymile country was a whole lot more than ten acres on a high ridge. The column was visible as we passed the settlement Big Delta, still seventy miles out. When we pulled in over the fire, it was about sixty acres and supercharged by a crown fire raging in seventy-five- to one-hundred-foot white spruce. The smoke lofted above us ten thousand feet. There were nine of us on the plane.

Eight would jump; the Redding jumper with the injured back would return to Fairbanks.

The six-man load on *Jump 19* had already jumped and were making their way to the fire. Erik the Blak would be fire boss. We winged in low for a closer look. The smoke column shadowed a large expanse of the country to the east. Luckily, the heavy timber, where the fire burned hottest, grew only on top of the ridge and along the south slopes. The north slope contained scrub timber and scattered brush, less combustible than thick white spruce. We would jump to the west, a thousand feet from the fire.

Oly came crawling back over the jumpers as *Jump 17* climbed to streamer altitude. He opened the door.

"You see where those other guys jumped?" he asked, yelling at the hole in the side of my helmet.

I nodded.

"We'll go the same way."

Oly put on his headset and began talking with the pilot. I passed out radios to Clouser and Seiler. I tucked mine into my right leg pocket and secured it with the tie straps.

Oly threw streamers, and I watched them drift up the canyon toward the fire. They were moving fast. He turned to me.

"Let's go with four squares the first pass. I'll drop the Forest Service guys last."

The three jumpers behind me got the idea and began getting ready. Oly looked at me. "Are you ready?" I nodded.

"Get in the door," he yelled. Completing my four-point check, I looked out directly at the fire. People often ask, "Aren't you afraid of jumping into a fire?" I explain that we don't actually jump into fires, only close by. Then they ask about being sucked into the fire. We recognize that, too, and take careful measure of the winds. There are times, though, when flying directly over a fire and looking straight into a massive smoke column, that the impression is so strong that it causes me to think that, this time, I *am* jumping into a forest fire. That's what it looked on the Fortymile jump.

"Stay downhill," Oly yelled, at the four in the first stick. "Away from the ridgetop. Watch the draft down low. I'll carry you plenty long."

Jump 17 came around on final as I braced myself in the door, looking directly into the column. Seconds passed. Then Oly pulled his head back.

"Get ready."

The slap came and in the next instant I was falling under the belly of *Jump 17*. Along came Kubichek, Clouser, and Seiler right behind. I opened, then turned and watched them do the same. As soon as I had a full canopy, I went through my postopening checks, then grabbed my left steering toggle, pulled it down, and brought the parachute around to face the fire.

As impressive as the fire had been from the plane, it was even more so in the open air. From the plane I could see it, smell it. Now, hanging out over the canyon, I could *hear* it two thousand feet below, rumbling and roaring.

Of the four jumpers in the first stick, I was the closest to the fire. I checked out the spot. It looked straightforward—steep, scattered spruce. I decided that, although I would keep off the ridge, I would stay as high upslope as was safe, thus giving the others more options to find a way in.

Checking out my jump partners, I found Kubi and Clouser to be in good position but not Seiler. He'd been carried too far and was racing back with the wind toward the jump spot. I reefed down on my left steering toggle and turned left up the hill even farther so Kubi and Clouser could give Al all the room he needed. Al was coming on at an alarming rate of speed and fast running out of altitude. He had one last spur ridge to clear to make the spot. The problem was that he needed not only to make the spot, but he also needed time to turn back and land into the wind. Landing with the wind, he would hit at a minimum of thirty-five miles per hour.

I was on final into the upper reaches of the jump spot when Al barely cleared the last ridge, the shadow of his parachute racing up to almost meet him. He had cleared the ridge all right but was too low to turn into the wind. It was ugly, but I couldn't afford to watch any longer. I had my own hillside racing up.

At thirty feet I pulled both steering toggles down to belt level, caught a little upslope breeze, and effected a flare that perfectly matched the contour of the slope. I landed gently and tumbled easily off my feet. Once unsuited, I heard the guys yelling down the hill. Strangely, there was something else. What was it? Sounded like a big sea lion.

A quarter mile up the ridge, Erik the Blak was lining out his jumpers. Lustful enough to send bears packing, loud enough to stampede caribou, and forceful enough to return aliens to their respective spaceships, the Blak's voice is one of the unlisted wonders of the world. Whatever his chosen volume—usually high—the Blak vocalizes a rather bizarre union of passion, assuagement, intelligence, and capability. Direct and unpretentious, his voice gives wings to the word *intonation*.

I yelled downhill to see if Al was OK. A lead jumper's first responsibility is to see that everyone has landed safely before the airplane leaves with the emergency trauma kit and we are unable to contact the outside.

"Anybody see Seiler?" Clouser yelled.

I tried Al on my radio. Nothing. I felt a sense of urgency. Arden Davis, had hanged to death in his shroud lines before he was found. The crew moved downhill hunting for Al. I headed down, too, then word drifted up. They had found him. Al was OK. I had my doubts. For him not to be OK, he would have to be dead. I hurried down.

When I got there, Clouser and Kubi were pulling a big spruce off of Al. I ran up, knelt beside him, carefully lifted his face mask, undid his chin strap, and slipped off his helmet.

"Al," I said loudly. "Are you OK?"

"No," he moaned, closing his eyes, resting his head back on the tundra. "But I'm fast."

I loosened the jacket collar around his neck. Kubi leaned over making a shadow that blocked the sun from his eyes.

"Al," I said. "Open your eyes. I want to look at your eyes."

I checked the pupils. They weren't dilated and both were the same size—a good sign. His lips were gray, though, and trembling slightly.

"Count backward from ten for me, buddy."

"I don't need to. I told you, I'm not hurt. I just didn't want to land south of that last . . . spur ridge," he said, shaking his head. "It's a long way . . . the spot . . . the spot. I ran downwind. By the time I cleared . . . that last ridge . . . I was too low . . . to turn. So I took things . . . straight on . . . Ha, ha. Pretty much as they came . . . fast . . . Ha, ha."

The main "thing" that came was a fourteen-inch-diameter black spruce, the impact occurring square on the trunk, six feet off the ground. Shreds of bark hung from Al's face mask, and his suit had some impressive green skid marks across the front. The impact must have been near thirty miles per hour, maybe more. Al had ricocheted past and as he did his chute caught the top of the tree and pulled it over. Al was yanked back underneath just before it landed on him.

"When I came to . . . I couldn't move. I felt heavy. The tree had me pinned. I didn't know what'd happened. I kept trying to push the tree off. Then I got sick and felt like I was going to pass out again. Then I heard somebody yelling . . . but couldn't answer."

Al looked around, reached up and pinched a point just above his nose, then let out a small moan.

"I remember lying there and feeling like I might pass out and not wake up . . . Ha! That was strange. Then I thought, damn, I've got to get up. The rest of the guys will be doing all the work . . . They'll think I'm off somewhere goofing off."

We had Al unsuited by the time he'd finished his story. He reached down and began massaging his knee. Clouser reached down and pulled on his right leg. Al jerked it back and rolled over and got to his knees.

"I'm OK," he insisted.

"Take it easy," I said. "We need to check your legs out."

"Just help me up," he said, reaching out a hand.

I took his hand and helped him to his feet. Al took a deep breath and wiped the hair from his forehead.

"I told you. I'm all right. Now let's put some fire out."

I looked him straight in the eye, and he glared back. I've always re-spected Al a great deal but doubted he could be relied on to judge his own condition. I thought about trying to talk him into taking it easy, but asking Al Seiler to take it easy would have been like asking Attila the Hun to try on a tutu.

"I'll make you a deal, Al. Walk around in a circle a couple times. If you pass inspection, we'll go fight fire."

Al walked out about fifteen feet, turned, and came back.

"Don't worry about it. I'm fine." At that he grabbed up his PG bag, slung it over his shoulder, and started up the hill.

A photo of Seiler on a fire hangs in the standby shack. Shirtless, weighted down with web gear, a saw pack, three cans of gas, and a chain saw slung over his right shoulder, he brandishes a Pulaski gripped halfway to the head. Al's face is black with ash. The caption beneath plays on a comment made by an American soldier during the liberation of Europe during World War II. It reads:

WE ARE THE ALASKA SMOKEJUMPERS, AND THIS IS AS FAR AS THE BASTARD IS GOING.

Standing there on the hillside, I looked at Kubi and Clouser and shrugged my shoulders. Whatever Al's condition, I might as well argue with the stars.

Erik the Blak called me on the radio. He was scouting the fire, doing his sea lion number.

"Aagghhh," he growled. "When you get done monkeying around down there, you'll find a red flag line near the top of the jump spot." Quickly I turned down the volume.

"Hump your backsides up the ridge, and you'll come to where we're cutting a helispot. Don't be bashful about food and water—bring enough for twenty-four hours. We're shorthanded and won't be leaving the line once we're committed." Classic Blak—computer-brain logged on, filing details, forming his plan, setting his logistics straight, prioritizing naturally.

I first met Erik the Blak in 1973 when he was Eric Schoenfeld, jumping out of La Grande, Oregon. Now, after twenty seasons jumping with both the Forest Service and the BLM, he had become the lead spotter for the Alaska Smokejumpers. Upon graduating from high school, he had attended Reed College on a math scholarship, having scored eight hundred out of eight hundred on the math portion of the SAT.

"Aagghhh, that was a cinch," he once told me. "I finished early and left."

"Did you like math?"

"Not particularly."

"Then why did you major in it?"

"Aagghhh, easy!" he replied. "No homework. No term papers. More time to ski."

"Ski?"

"College was a practical means to support my skiing habit. They paid all expenses at Reed, and I took math exams for students at other colleges to earn extra money for lift tickets."

Midway in telling this, the Blak had stopped and stared vacantly into space.

"All that silly scholastic stuff I did in the library between classes."

After college Erik spent four years in the U.S. Air Force and was stationed in Canada. After Canada, he and his new wife returned to Oregon where he began smokejumping. At Cave Junction, the jumpers became so enamored with Eric's unorthodox character and sordid black felt hat that they nicknamed him Eric the Black. In the years that followed, Eric became Erik, and the Black became the Blak, and finally he became Erik the Blak. For brevity's sake, some jumpers call him the Blak. Two years ago the Blak spray-painted his jumpsuit and helmet black, which soon faded and left him jumping in a rat-colored outfit. The parachutes he rigs are signed BLAK PAK.

I've watched the Blak emerge from his office to meet officials from Washington, D.C., private aircraft contractors, aeronautical engineers, and dignitaries from foreign countries. They arrive in official uniform, suit and tie, bearing briefcases. Wearing patched "Can't Bust 'Ems," a tattered flannel shirt, and his infamous black hat, the Blak accosts them, his upper lip wedged full of Copenhagen, some of it stuck to his teeth.

With these astonished visitors, the Blak discusses any and all manner of aircraft contracts, aircraft flight characteristics, general aeronautics, gravitation physics, or for that matter, anything even remotely pertaining to smokejumping. At the blackboard, he draws zany diagrams of air flowing over the wings of aircraft that look like flying fish. Along the margins, mumbling to himself, he scratches barely legible Federal Aviations regulations, contract clauses, subclauses, and baffling equations. When it comes to aircraft and what makes them fly, few people can match such rare genius. Over the years, the Blak has, no doubt, saved hundreds of thousands of government dollars, and quite possibly a life or two.

In one trip, the eight of us packed all we could to the top of the ridge. It was midafternoon, hot and humid, and we were sweating bigtime before we reached a small flat where two of the Blak's snookies were cutting a helispot. Up the ridge another quarter of a mile, we came to the fire's edge. Out of the smoke materialized a werewolf who had found himself a hard hat. Immediately he went for his radio, telescoped out the antenna, and turned south toward Tok Junction.

"Look," the Blak said, exasperated, "I'm trying to catch a fire here. You'll get the rest of our order later. I've got a bunch of high-octane brush apes out here gathering up their toys, and we're going to work. I'll call you again at midnight. Grrrrrr."

The Blak's plan included an abundance of grunts and growls. I was to take ten jumpers and one chain saw and go around the troublesome right side of the fire. He would scout the left. We would both move east along the ridge into the night and try to cut off the head before the heat of the following day.

Al cranked up a chain saw and dove into the work as if it were some kind of tonic. We began clearing a fire line along the right flank, Al cutting out the brush and downed logs, and the rest using our Pulaskis to scrape away the forest litter down to mineral soil. A retardant plane came roaring in low and dropped ahead of us along the ridge. Around eight o'clock I reconned east in the direction of the head, listening over the radio as the Blak talked with Rob Collins in the Air Attack ship.

"Ahhgg," said the Blak. "Problem is, it's slow going in the timber. We're short of bodies. I doubt the retardant will hold until we get there, but drop it anyway. It's our only chance."

The rest of my group continued building line along the fire's edge as it fingered up and down the steep sidehill in tall white spruce. The fire was outdistancing us by too great a margin. It would slow down at some point during the night, but at its current rate of spread, even with the slowdown, it would grow too big for us to contain before the hot part of the next day. The fire was burning hottest nearer the top of the ridge, and there the line had to be dug wider and more trees needed to be felled. This further slowed progress. I continued scouting the ridge, looking for natural barriers, a water source—anything to serve as inspiration for my line builders. Not that they needed it. When Al first stopped to refuel his saw, one of them asked about his "wreck." Sweat-soaked and pumped up, he brushed off the question and pushed on, driving himself through one tank after another.

Around 9:30, Rob Collins came back on the radio.

"OK, Blak, we're going to leave you now. We put the last two loads across the head and it's lying down. More retardant won't help if you don't have people to take advantage of it. They've got another fire for us, ahhh . . . your fire stays pretty much on the ridge with a few fingers down the right flank. The left flank looks to be no problem—for tonight anyway."

The Blak growled information to be relayed to Tok Forestry, the state headquarters with management authority over our fire.

"All in all, Erik, I think you've got a fair chance at it tonight. If you get some help in the morning you may be able to hold it. If not, we'll have to order some crews to help protect Chicken. There's nothing out in front of your fire but miles and miles of white spruce."

Collins wished us good luck then flew off to the north, taking the last air tanker with him.

At 10:30, I reconned the entire head, staying on top of the ridge and cutting through the burn where possible. The head of the fire was five acres or so—about the size of five football fields. Just as I arrived, things started heating up. The fire retardant was rapidly losing its effectiveness. I tried scratching line, but my efforts were puny compared to what was needed. Suddenly, in a great swoosh, a big spruce torched, sending flames a hundred feet into the air, then another tree ignited, and another, until the head was building toward a crown fire. Crackles and hisses issued forth from snags and large downed logs. The wind began to draft upslope. Sparks rained down in front of the head. I called the Blak. He hiked to a high point and was able to reach Tok Forestry on the radio. "We need retardant," he snarled at them. A small scratchy voice answered plaintively, but I couldn't hear what was said.

"Aaagghh!" The Blak cut in. "I don't know if we'll get it either, but I'm ordering it anyway." Another small scratchy transmission.

"Do it any way you like," the Blak growled. "But if you've got any friends living in Chicken, you better send us some mud tonight."

At midnight, *Tanker 124* out of Tanacross, a PB4Y carrying twenty-four hundred gallons, arrived and dropped directly on the head. As *124* cleared the area, Air Attack arrived from Fairbanks in lead plane *47 Alpha*. *Tanker 84* was close behind.

"With one more load of retardant, we can most likely hold this," I told them. "If not, we're probably screwed."

Air Attack called Tok and asked to have *Tanker 124* load and return. The small scratchy voice muttered something about *Tanker 124* having had a rough day and that they were running low on duty time.

While *Tanker 84* made its final runs, I dug line as hard as I could, rounding the head on the left flank of the ridge. Knowing how Al and the other line builders were busting butt motivated me to give it my all. Zooming low over the trees and roaring out over the canyons, *Tanker 84* finished its last drops. With no crown fire, the head cooled down. I radioed my crew and urged them to spread out and move up.

Just after one in the morning, *T-124* showed up and put another two thousand gallons in the timber across the remainder of the head. As soon as they left, we moved up into the heavy smoke, connected our two sections of line, then spread out again, patrolling back and forth, knocking down flare-ups, and falling snags.

Eleven of us worked the right flank as light began gathering in the east. The Blak had reconned the rest of the fire and firmed up his plan for the coming day. The left flank had burned all the way to a creek and would hold.

By 3:00 that morning we'd pulled the bigger chunks of logs and limbs back inside the burn and placed them so they couldn't roll. At 3:30, four of us took chain saws and cut an emergency helispot on top of the ridge about two hundred yards east of the head. By that time the first helispot was a mile and a half back. The second helispot would be strategic when the crews began arriving.

At 4:30, I rounded everybody up. Exhausted, we gathered on a grassy area and ate the food we'd brought from the jump spot. Kubichek hauled out what was left of his chicken and finished it off. Deciding on some sleep, we took out our jackets, raincoats, and tarps. Coyote camping, or just plain "coyotin'," is what we call it. For warmth, you put on the extra clothes stashed in your PG bag, then spread out your tarp and roll up in it.

The Blak called on his radio.

"Aaagghh," he growled softly. "Good job down that way. Try to get some rest. Tomorrow's gonna be a long day. Good night . . . grrrrr."

A rough-looking bunch stirred that morning at 7:30. I woke dry-mouthed and groggy. The others were sitting quietly, staring straight ahead at nothing. After a while they rubbed their eyes, scratched their whiskers, drank from canteens, shook the cobwebs from their heads, then got to their feet and began organizing their gear.

Tok Forestry's plan was to replace us that morning. There was a statewide priority on getting jumpers back into jump status; replacement crews were ready to be flown in. The new fire boss, a state fellow named Abrahamson, was coming in to take over the fire from the Blak. Meanwhile, we would eat, then spread out and keep working the line until the crews arrived.

We finished most of our food, drank coffee, talked about the fire and what still needed to be done, and speculated about the weather of the coming afternoon. Warm air began flowing out of the canyon to the south and over the ridge. We were beginning to feel more alive. The food picked up our energy, and our spirits were soon lifted by the easy laughter that seems to follow in the wake of extreme effort.

I kept an eye on Al. He hadn't said much, but that wasn't unusual for a man who gets a great deal of pleasure out of being with others and just listening. He'd eaten a big can of beans, another of fruit cocktail, and he seemed to walk all right. I knew that he could be hurting and not say a word. Al's nature is to pass his troubles off to something, but I'm not sure what. Maybe it's work and just keeping quiet. I'd seen him do it before, work like a maniac, stay out on the fire line after everyone else had come to camp, driving himself with a relentlessness that I'd feared might someday break him.

After a moment of circumspection, I walked over, sat beside him, and spoke in a voice careful not to draw attention his way.

"Hey, Al," I said. "Good job with the saw. How's your head?"

"Don't be worrying about my head," he came back, glancing quickly over at me then down to the ground. "There's not enough in it to matter anyway." After a small pause he looked at me again. "Really . . . I feel fine."

"You always feel fine, but head injuries can be dangerous. We can't be losing one of our best now, you hear?"

"Well . . . you sure won't be losing much if you lose me."

"That's bullshit and you know it. Everybody knows it. I don't like the idea that I saw you hurt and didn't do anything about it. And that goes for the rest of you, too," I said, turning my comment to the whole group. "We've done well on this fire, but we're getting tired. Watch out for each other. It's times like this when we tend to let our guards down."

"Listen to Old Leathersack, boys." Al chuckled. "He's got enough knots on his head to know what he's talking about. It won't be long before the snags start coming down."

I looked at Al. He was grinning. Maybe he'd appreciated my asking about him after all. In his own way, I think he was letting me know that he respected what I'd said.

Besides the dent in Al's head, one of the McCall jumpers had slashed his left calf with the chain saw, which left three ugly finger-length strips of flesh hanging just above the top of his boot. He'd stuck the strips back in place and wrapped the area tightly with strips torn off his T-shirt. He claimed to be fine, too, even though the bandages were soaked with blood.

In spite of our difficulties, we were in good spirits. A nasty 150-acre crown fire in thick white spruce had been—for the time at least—contained. We'd done it with only fourteen jumpers, the scratchy little voice from Tok, and some excellent help from the air tankers. In terms of the obscure and mostly unseen triumphs that come to wildland fire-fighters, what we'd done there in the mountains of the Fortymile was typical. As an initial attack force, we had caught the fire. Maybe we could hold it.

After gathering our gear we spread out down the line to work the perimeter, patrolling, falling snags, and waiting for word about our replacements.

At 11:30, it came. "Ahhgg," bellowed the Blak. "Chopper's down. They got a warning light—had to put her down in the weeds between here and Tok. We're here until they find a mechanic or another bucket of bolts to fly us out." That bit of logistics took the rest of the day. The winds remained calm, and the hours passed peacefully, with only a few midafternoon flare-ups.

At nine that night we walked the long ridge back toward the tail of the fire. The helicopter had been fixed. Abrahamson had arrived. New crews were being flown in but too late to do a night shift. There was constant talk on the radio. Plan B was for the jumpers to leave first thing in the morning.

When we arrived at the tail, fire camp was being set up in a half-acre saddle under a stand of mature paper birch. The woods were filled with Alaskan natives. Most of Alaska's seventy-five EFF crews are either Eskimo or Athabascan Indian, sometimes both. Each year they leave their villages for up to a month at a time to fight fire. It's been a tradition for forty years, and most of their crews are highly respected in Alaska and the Lower 48.

We walked over to where the Blak and the two snookies stood talking by a big pile of smokejumper gear. To our extreme relief, they had retrieved all the jump gear and paracargo from the jump spot to the top of the ridge. The three of them had lugged some two thousand pounds of equipment up a damn steep mountain in the hot afternoon sun, a task that had taken them four hours. All the way to camp I'd been thinking to myself, I'm dead-assed tired, and I still have to go down that steep-assed hill and pack all that heavy-assed gear to the top of this high-assed ridge.

I pulled off my PG bag, dropped it heavily to the ground, lay down, rested my head on it, closed my eyes, and listened for a moment.

The camp was a bustle of activity as the members of the Mentasta, Dry Creek, and Tanacross EFF crews set up their tents, strung tarps, dug fire pits, chopped wood, laughed, and filled the air with the sounds of men and women working to make a home for themselves in the outdoors. People scratched in the leaves making places for their sleeping bags. Bird whistles and animal calls were traded back and forth.

I opened my eyes. A light breeze moved through the saddle. Wildwood roses bloomed close to where I lay, their pink blossoms glowing in low-angled light. I had begun to stiffen up. Leg weary, I got to my feet and began what I hoped would be my last task of a very long day. In short order, the other jumpers and I had our tents up and two cooking fires going. Fat boy boxes and cubies of water had been placed nearby, and buckets of water were boiling. We fixed dinner and ate, but no one hung around to tell jump stories. By eleven o'clock we were all down, even though half the camp was still up.

Shortly after I crawled inside my tent and got comfortable, I began to experience a common but unwelcome distress. After five nights of little sleep, the most annoying consequence of chronic sleep deprivation had finally set in. Exhausted, I lay there wide awake, my mind racing. Tossing and turning, I was being held captive to a series of images filled with running spot fires, diving air tankers, and Al lying unconscious under a tree.

Relax, I said to myself. *Try to visualize a calm mountain lake, a meadow of waving grass.* Nothing worked. Desperate, I tried praying. Sincere at first, I started by thanking God that Al was OK but then began to question if he really was, and soon lost my focus and found myself contemplating some trivial nonsense about Big Ernie. Sleep time was precious and I was wasting it. Exasperated, I took out my watch. It was 12:45. Using the second hand, I measured my pulse, estimating it at a hundred—thirty-five beats above normal.

The next day we arrived in Tanacross, whereupon our silver-tongued sea lion sweet-talked the local dispatcher into getting us a hot meal. An hour later, we were in Tok Junction at Fast Eddie's ordering hamburgers, french fries, pizza, chili beans, steaks, salads, milk shakes, and apple pie. I had gone to the restroom and tried to wash off five days of dirt. I felt rough, but at the same time relieved to be back in an environment of convenience. It felt good to walk on flat ground, sit in a chair, and eat off a table. While the others had waited for the food, I stepped outside to a telephone booth. The line rang quite a few times, then a man answered.

"Is Sally there?"

A lengthy pause. I started to ask again.

"Just a minute."

"Hello?"

"Sally, this is Murry."

"Murry," she said. "I thought you'd been eaten by a bear."

She told me she'd read about the fires but didn't have time to talk. She was going to be late for work.

"Are you doing anything later tonight?" I asked awkwardly.

"Just working."

"I know you're busy, but I was wondering if I could see you. Maybe just for a drink or something?"

"I'll have to think about it." A small silence followed. I waited.

Think about it? Think about what? Why is it that women always have to think about it? There wasn't time *to think about stuff.* I waited, frustrated, trying to think of something sensible to say.

"Call me at work around ten. I'll know then what time I get off."

I went back inside and tore into the hot, fragrant food. We ate like hyenas at a night kill, cramming food down our throats as one might stuff a turkey. Everyone else's food looked better than mine. Smelled good. Looked good. I wanted it all. The food in the fat boy boxes will keep you alive, true, even sustain a certain level of energy, but when output exceeds intake, pounds drop off. First the fat reserves burn up, then the muscles start to go. That's when the craving for fresh food becomes acute. Filthy and wild-looking, we wolfed our dinner down shamelessly, while tourists seated nearby cast nervous glances in our direction.

Suddenly Seiler jumped up and ran outside holding his hand over his mouth. Clearing the door, he projectile-vomited in between two cars right at the edge of the sidewalk. In a few minutes he came back but went off by himself and sat at a table alone with his head cradled in his hands.

Our flight back to Fairbanks was routine. I lay on a pile of gear, watching the Alaska Range pass by and thinking how stupid I'd been to call Sally. I had no idea what was in store once we got back to town. Was it need, or did I really like her? And, perhaps more important, could I tell the difference? In many ways it had been like that with Steffani.

"Shit," I mumbled. "My life's a fucking mess."

I met Steffani in the early eighties. The daughter of Wyoming ranching blood, she'd been raised in Puerto Rico and took to the sea in her late teens, sailing into the Caribbean, then west through the

Panama Canal into the blue Pacific, and on to California. Our first months together were an enchantment of candlelit saunas and starry winter nights making love in her tepee by the light of a campfire. After two years of passion and catfights, our relationship had progressively deteriorated into a scary cycle of sex and war—two good people in the wrong relationship. Steffani moved out and set up housekeeping three miles down the road, but we couldn't resist the temptation to see each other. I'd spend all day hauling firewood to her new place and stacking it just like she wanted it on her back porch. Toward dark she'd invite me for dinner. At some point in the evening, just about the time I'd begin to think that we were finally going to make up, we'd find ourselves not agreeing on some ridiculous new-age issue concerning low self-esteem or co-dependence, and then all hell would break loose. "Just fucking stop it," she'd scream. "Why does it always have to be your way? Your plans, your dreams, and that endless stupid need of yours to keep jumping out of airplanes into fires? I can't stand it anymore. Get out of here, and let me live my life in peace."

I'd leave stunned and bewildered but firmly resolved to never see her again. Down the road, I'd pull over for a couple minutes, then turn around and go back. Steffani would be waiting at the door, dressed in her nightshirt, crying. We'd grab each other and kiss wildly and wrestle our way to her bed where we'd proceed to make some of the best love I've ever had.

It was a dreadful situation. Each time it happened I became more frightened. Finally Steffani had a bold affair right in front of me, right where I couldn't help but see it. Each morning, in the soft light of dawn, I'd drive by her house on the way to my horse-logging operation and see his car in her yard. Maybe that was the only way she could think of ending the mess we'd gotten ourselves into. I don't know. Whatever she thought, the whole thing left me heartbroken.

In 1986 I became involved with Robin, a six-foot-two, shapely radiologist with auburn curls, who had an extremely healthy sexual appetite and a bounding, fun-loving Airedale named Mariah. During the summer of 1988, while Yellowstone was burning and I was earning a thousand hours of overtime fighting fire in Montana, tall, beautiful Robin was having an affair with a mutual friend. After one of the most demanding smokejumper seasons ever, I finally got home in early October battered and beaten, a complete wreck, and walked right into that.

By 1990, in self-defense, I was spending a lot of time by myself reading books like *Loneliness and Love, Some Men Are More Perfect Than Others, Witchcraft and Psychiatry,* and an especially vexing one entitled *Love Addict at Eighty-Four: Confessions of an Old Romantic.*

Fifty miles southeast of Fairbanks the sky disappeared in a gray-brown haze and the sun faded dull orange. Twenty miles out, at a thousand feet, I could barely see the ground. Like a pall of disorder, the smoke got thicker and thicker until its scent was heavy in the plane and the ground completely disappeared. Everyone else in the plane slept. I was convinced that Al's puking had resulted from a head injury.

As we pulled up to the standby shack, the ramp out front looked like a used-car lot for airplanes. Several had arrived from the Lower 48. There they sat, some of ours, some of theirs: *Jump 19, Jump 07,* a Missoula Twin Otter, Missoula's Turbine DC-3, and McCall's, too. All lined up in front of the shack with not a propeller turning.

Inside the ready room we found out that all air operations had been placed on a restricted basis. There were fires, dozens of them, but the smoke made flying too risky. Down the ramp, parked in blue haze, two acres of air tankers also sat waiting.

"Taylor," the man at the operations desk said, pointing over his shoulder at a notice on the bulletin board. "How long since you've had a day off?" I glanced at the note: IF YOU'RE PAST TWENTY-ONE, SEE US ABOUT A DAY OFF.

I pulled a notebook out of my shirt pocket and started flipping through it.

"Can't remember, can you?" the operations man said. "Now's the time to take it. We're not flying, so you won't miss anything. Besides, they're really pushing this. Everyone's beat."

"Forty-three days," I said.

"You're outta here," he cried, sliding my name tag out of the jump lineup. At first I was miffed, but then I remembered Sally.

"Be back here at 0800, day after tomorrow. Take a break, man. You need it."

When I picked up Sally, she bounced out of the Captain Bartlett Inn wearing tight Levi's and a sheer white blouse with a negligee-like neckline.

"I wanted to be comfortable," she said, twirling around, "so I brought this to work with me."

However good Sally had looked when I'd last seen her the week before, eating clams on the deck at the Chena Pump House, she looked twice as good now. *Good* good, as in woman, good as in sexy, all 104 pounds.

We drove west out Airport Road on our way to the Chena Pump House again. Suddenly all felt right with the world. I was with Sally. And I was going to an establishment with fine rugs and paintings, to sit like a gentleman with a lady, instead of running around chasing fire all night.

Inside the Pump House, Sally and I took a seat in a corner, sipped our imported beers, and ordered a large plate of steamed clams. Earth-tone tapestries hung on the walls between picture windows framing views of the river. Chandeliers glittered above a rich darkness of antique chairs and tables. Our own chairs were rounded velvet antiques with splayed claw feet and eagle-talon arms. Sally sat upright like a queen, alternately casting a fond eye over the room, me, and the bowl of clams.

"You look tired," she said. "Was it a bad fire?"

"We had three. All in beautiful places. Actually, none of them were that bad, but three fires in six days is tough no matter what."

"You've lost weight, too," she said, sending me a concerned look. "My boss at the inn was talking about all the smoke and how it will ruin Alaska for the tourists. It seemed like a funny thing to say. I told him that everything possible was being done, but he complained that too many fires were being allowed to burn. If it keeps up, he says he'll have to lay some people off."

"The let-burn policy has been temporarily rescinded," I said. "No one thought it would stay this dry. Tell him that these are record conditions and that once fires get to a certain size there's nothing anyone can do until the rains come. In the meantime, as soon as the smoke clears so we can fly again, we'll be doing our best to put out all the new ones."

"That's so exciting," Sally said. "You have a wonderful job . . . I mean it's important—more important than tourism. I wish I had work that was interesting."

She folded her hands on the table, consulted them briefly, then sighed.

"I want an adventure in my life," Sally declared emphatically. "I'm tired of waiting tables. I want to see Alaska—the real Alaska—explore it, dream in it."

I mentioned that I had the next day off.

"So do I." Sally beamed.

"Let's do something."

"It'll be our first full day together," she said. "We can celebrate."

As I drove Sally home, taking the Steese Highway to Four Corners, I recalled when I'd looked down three days before during the plane ride to the Fortymile fire. It had been only fifty-six hours, but it felt like two weeks. I turned down the dirt lane to her brother's cabin, and pulled into the yard in front of the same green-roofed garage I'd spotted as we flew over.

Starting to get out, I suddenly reversed direction, leaned over, and kissed Sally on the cheek. She didn't move, so I kissed her again, still on the cheek, but more fully. She turned a little, and I kissed her lightly at the corner of her mouth. Taking her in my arms, I leaned back against the seat, content just to be there holding her and smelling her hair. After a while I involuntarily let out a strange groan. Sally cocked her head and pushed against me, pinning me behind the steering wheel as I cradled her against my right side. She pushed a tape in the cassette player and turned the volume down low. We snuggled for a while, me running my fingertips up and down her forearm.

"You're tired," she finally said. "You've been tired all night. I could tell it the minute I first saw you." I nuzzled the side of her head, wanting another kiss.

"Why don't you come in? There's a bed upstairs, and Mike's gone for the weekend. You're in no condition to be on the road."

Sally and I walked up the wooded path to the cabin. The porch was spacious and littered with wood chips and kindling. Inside, the cabin was dark and cool, and smelled of woodsmoke. We negotiated the stairs and found the extra bed. I kicked off my shoes and lay back on the pillow. The last thing I remember was Sally covering me with a big green quilt.

Bleary-eyed I looked around the dim room. I wasn't on a fire but in a strange house somewhere. What was I doing? What time was it? On the floor beside the bed were my shoes, and then I remembered. Sally.

I started to get up but immediately lay back, heavy, stiff, and sore. Thick-tongued and rough, I understood the plight of poor Frankenstein's monster, a man hastily assembled, then abandoned at dawn. My left knee felt like someone had hit it with a hammer. My eyes were dry; I needed water. Veteran jumpers have been heard to say, "It's tough to push jumping into your fifties." Maybe I was giving out and didn't know it.

A door squeaked, then gently shut. God! I hoped it was Sally and not her brother, come home to find me upstairs in his house, in his bed, and, as far as he could tell, possibly with his sister.

The water started in the shower. The sound was soothing. I imagined Sally just inches away on the other side of the wall, wet, steamy, and pink. I imagined her washing her hair, neck, arms, breasts. Images of her dripping with bubbles appeared and disappeared in steam. Her large eyes were closed as she reached down to wash between her legs. Turning, she faced the other way, reaching around behind her neck with a soapy rag.

The water went off. I listened closely. The creaking door again, soft footfall in the hall. My door opened slightly. Sally peeked around the corner, wearing a plastic shower cap and a white terry-cloth robe.

"Good morning," she whispered, removing her cap and drying the back of her head. "Did I wake you?"

I didn't answer right away, my mind being tripped up by what I was seeing partly exposed in the deep V-ed front of her robe.

"Oh . . . no! Not exactly . . . You look . . . nice."

Sally smiled, ruffled her hair with her fingers, and disappeared around the corner down the hall.

"I know a great place for breakfast," she announced. "You interested?"

After breakfast, we spent the rest of the morning at Pike's Landing, watching the crews load tool kits, survival gear, high-powered rifles, and emergency rations into their powerboats. The Yukon 800 is a marathon race down the Chena to the Tanana, down the Tanana to the Yukon River, and then west through the middle of Alaska's vast interior. The first leg ends at Galena, four hundred miles downriver. Racers spend the night there resting and making repairs, then return the following day. First run in the fifties, the race has become the ultimate challenge for the members of the Fairbanks Powerboat Association.

We sat on a grassy area above the beach awhile until it got too warm, then moved inside and sat at the bar watching the crowd out the big bay windows. Feeling sleepy, I asked Sally if she minded if we went home. We drove to her brother's place and parked in the yard and started kissing.

"You can take a nap here," Sally said. "Come on."

We went upstairs where I'd slept the night before. While I stretched out on the bed, she went around to the windows and opened them.

"Thank you," I said. The last thing I heard was the door closing softly.

When I woke a couple hours later, I found myself covered by the big green quilt. Trying to move, I discovered something beside me.

"You kept moaning and mumbling," Sally said, "so I'd come and sit on the bed beside you and you'd stop. After a while it just seemed like a nice idea to join you."

I took Sally in my arms and began kissing her. When I lifted up the quilt to get closer, I saw that she wore only a Mickey Mouse T-shirt and a pair of black panties. Mickey's wizard hat sat tipped jauntily to one side, while over his head he waved a wand trailing stardust.

12

The standby shack was a madhouse when I arrived the next morning. Jumpers crowded into the ready room, greasing their boots and talking as they waited for roll call. We were starting an hour early. I checked the jump list. A total of 235 smokejumpers—more than half the number in the entire United States—were now in Alaska. The Alaska Fire Service had ordered 170 jumpers north from the Lower 48 to complement our 65. A ten-man parachute-rigging crew had been ordered as well. They were rigging on average 120 parachutes each fourteen-hour shift. Since we had not flown the day before, our parachute storage shelves were looking fairly full.

Fifty regular jumpers had regrouped back in Fairbanks; the rest were still out on fires. An eight-man load stood by in Galena. Another dozen were tied up in operations, logistics, and paracargo. A few had been injured. The locker area looked like a disaster relief facility, with travel bags stuffed under benches and stacked along walls. Spruce cones, twigs, small branches, leaves, moss, gum wrappers, bottles of bug dope, used maps, dirty socks, and half-eaten plastic bags of dried fruit and jerky littered the floor. Spaces between lockers had been claimed by visiting jumpers and summarily converted into mini–hobo jungles. In my absence, one such group had set up camp next to my locker. Faces moved in and about, smiling, some familiar, some new. Eager handshakes and introductions were exchanged in a reunion celebrating old friendships and other fire busts. People laughed and yelled back and forth, repacking PG bags and inspecting jump gear.

"Roll call," Doug Swantner announced over the PA system. The jumpers crowded into the area between the suit-up racks and the ops desk.

"OK, listen up," Swani yelled. "The first load will be a ten-man load—Troop, Clouser, Taylor, Kubichek, Seiler, Nelson, Decoteau, Quacks, Robinson, and Bell. We're gonna be flying shortly, so be ready. We got a few items to cover this morning and not much time to do it, so listen up and we'll try to get through it . . . Boats?"

Tom Boatner, our crew supervisor, stepped in front of the ops desk, his face badly sunburned. He'd lost weight, and the hollows of his eye sockets were deeply recessed in the pale skin that had been covered by his goggles.

"We're catching almost everything we're jumping," Tom began. "Considering the conditions, that's damn good fire fighting. I, for one, want to say thank you for all the hard work. Hang in there. We know you're tired. If you need rest, let us know. We'll work something out. We've been jumping about one hundred fires a week for four weeks now . . ."

"*We?*" Leonard Wehking piped up, sending a chuckle through the group.

"Well," Boats said, looking sheepishly at the floor and shuffling his feet, "I was with you in spirit anyway. Good timing on my part, I'd say."

Boats hesitated a second, looked down at his notes again, then grinned.

"I guess I better tell you. After we summited McKinley—just over twenty thousand feet—we were all pretty bushed. But our homegrown hero here," he said, pointing his clipboard at Buck Nelson, "that wasn't enough for him. He got down and did thirty-five push-ups at the very top—*with* a fifty-pound pack on his back."

A hearty roar shook the shack. Hoots and hollers mingled with multifarious *ah—uhhhs,* chimpanzee wah-barks, and other primate sounds. Buck stepped forward, clinching his hands over his head in victory. Once we'd finished honoring manly men, Buck, Boats, and male hysteria in general, the group quieted, and Boats continued the briefing.

"We're really tight on radios—get them back to Fairbanks or be sure you turn them in to a spotter at an outstation. Try your best to back-haul your main chutes with you. Leave the fire packs and other replaceable gear, but we need the mains back here and also as many cargo chutes as you can squeeze in. If you've fire-bossed a fire and haven't got your fire report in yet, it's waiting for you at the box. Keep caught up if you can. This could go on for another month.

"Let the box know of any problems you're having. We'll see what we can do." Boats looked over at Swantner. "That's all I've got. Can you think of anything else?"

Just as Boats concluded the question, the phone rang. Swani picked it up, listened for a second, held his hand over the mouthpiece, and yelled out, "Hey, don't go away, we got something."

The group held up. I moved over to the jump list to confirm that I was on the first load. Swani hung up the phone and reached for the PA microphone. "I think we've got a fire call. They're running down the particulars up in the Puzzle Palace, but first load get ready. The pilots won't be here for ten minutes, but I think it's gonna be a go."

I suited up, grabbed my PG bag, and stepped out onto the ramp and sat on a bench. The morning air was strong with the smell of smoke, and the visibility was still poor. Three miles away I could barely make out the tops of a few tall buildings downtown. The skyline was a flat yellow-gray. The day was already warm—almost hot.

Suddenly there was a roar, and I turned to see one of the air tankers sitting in a cloud of smoke, its strobe lights flashing. While Jim Olson came over and put me through my buddy check, the air tanker moved onto the runway and began its takeoff roll, the roar of its four giant radial engines resonating mightily in and around all the buildings. At the same time, *Jump 17* began running up its twin turbines.

"Helmet, gloves, and letdown rope?" Oly yelled.

"Got 'em," I yelled back, nodding.

"Have a good one," he said, grinning and slapping me on the shoulder.

Jump 17 had no sooner cleared the runway than most of the jumpers began scrambling for comfortable places to sleep. I waited a little, then got to my knees and looked out the window. The Steese Highway ran north to Four Corners. My eyes followed the little dirt road through the birch woods and found two green roofs. Down there somewhere was a log cabin and a big green quilt. Sally's scent suddenly came to me so vividly that I found myself aroused by my own imaginings.

The plane angled to the east and leveled out, the left wing yawing forward and removing Four Corners and the green roofs from sight.

A note came back for Troop, who lay sprawled on the floor in the rear of the plane. I snagged it as it came by.

Fire B-405

8 acres in black spruce and caribou moss—100% active. High ridgetop—dry

Air Attack 264 responding—Fairbanks.

Air Tankers: Tanker 84—Fairbanks

Tanker 33—Fort Yukon

Tanker 124—Tanacross POSSIBLE REBURN!

I passed the note to Troop. He considered it at length and stuffed it inside the top flap of his PG bag. Closing his eyes, he laid his head back on his main parachute. The rest of the crew seemed to be sleeping peacefully amid all that noise. So peacefully, in fact, it looked like we were hauling a load of corpses. It was hard to do much sightseeing out the small windows of *Jump 17,* so I lay back and stared at the ceiling. Someone had scratched above one of the windows—THIS SURE DOESN'T LOOK LIKE KANSAS, TOTO.

I looked over our load. It was a strong one—Bell, Mitch, Nelson, everyone. Seiler lay sprawled out, head arched back, resting flat against the floor, mouth wide open. After the Fortymile fire and his puking at Fast Eddie's, some of the guys had mentioned that maybe he should get his head checked—an idea he wanted no part of. Somehow the base manager got wind of it and insisted he see a doctor.

"Yeah, right." Al had grumbled. "See a doctor. What'll they do? Run a bunch of tests and charge the government ten times what it's worth. By the time the results come back, fire season will be over. No thanks. My head's too hard to damage anyway."

Finally, Rodger put his foot down and told Al to fill out an injury report and consent to a physician's examination or be pulled off the jump list. A brain scan revealed that Al had suffered a moderate concussion but no more. So what Al had said all along—big bill, no damage—had been essentially correct. Just the same, it was a relief to know that he'd been thoroughly checked out and cleared to go, even if he did look like a recently expired mouth breather stretched out there on the floor.

Another note was passed back. The jumpers stirred in turn, passing it along, obviously not pleased with being disturbed.

Troop:
Will hold in Eagle. Not clear what's up. May not take action. Fire in Canada (B-405). Will stand by for B-499. Tankers have been turned around.

Troop took the note and looked at it just as he looks at all such notes—at length. Then, apparently resigned to our new fate, he withdrew into himself like a big turtle, staring blankly for a moment, then closing his kind brown eyes.

Heavy smoke drifted along the horizon in the east toward Canada, but for the most part, the Fortymile country was having a bright and beautiful day. We approached the airstrip from the river end. As *Jump 17* descended over the Yukon, we got a bird's-eye view of town.

I could see a line of small boats scattered along its muddy bank.

Boats are to people in the bush what cars and pickups are to people elsewhere. They serve as the transportation of choice for hunting and traveling between fishing camps, for gathering firewood and visiting neighbors—some of whom live hundreds of miles up and down the waterways.

Like a lot of settlements in Alaska, Eagle had come into being at the time of the Klondike gold rush of 1896. With barely a hundred residents, the town is both old and new. The old part is mostly log cabins and a few larger log structures tilted this way and that by the seasonal heavings of permafrost. Many are overgrown by willows. The newer portion contains frame-constructed homes with cleared yards. The two parts are connected by a network of foot-worn trails running through the grass and weeds. Near the center sits a New England–style church, its steeple radiant and white in the midst of a community isolated by wilderness.

After parking *Jump 17,* we off-loaded and spread our jump gear out in the shade of the wings. I privately thanked Big Ernie for sparing us, for the moment at least, from another fire. Some of the jumpers lay down on their gear and speculated about what might be in store for us; others took out books, read a few minutes, then dropped off to sleep.

Fifteen minutes later a pickup roared up, and the BLM station manager got out and shook hands with our spotter, John Gould.

"We've got a couple problems down here," he said. "Dalan Romero is on B-499 south of here with fifteen jumpers. They've been on it since Wednesday morning, so they're really done in. Dalan thinks they'll know by midafternoon if they can hold it. If the wind kicks up and things go bad, we'll send you guys out there."

"How far out are they?" Troop asked.

"Forty miles. We've got good radio contact with them," the manager said.

"The second problem is this fire you were running on, B-405. It was jumped by a couple jumpers six days ago. They ran out of food, had a bear problem, and had to leave it. It's right on the border. We're checking with the Canadians to see what kind of a joint commitment they're willing to make."

"How far out's that?" Troop asked.

"It's also about forty miles."

"Just let us know which one you want us on," Troop said. "We're not doing anybody any good sitting here."

"OK, sounds good. If you guys are still here at one o'clock, I'll have some lunches sent down."

We thanked him and he drove off, rattling down the road back to town. Feeling drowsy, I made a pillow by folding my fire shirt and placing it on top of my main chute. I lay back in the cool air, but the jumpers who had slept all the way from Fairbanks began talking about women.

"They're dangerous people," Mitch Decoteau claimed flatly. "Let me tell you."

"How's that?" Quacks mumbled.

Mitch lifted up on an elbow and glanced casually over his shoulder.

"Hey, Quacks," Mitch grunted. "If you have to ask, you probably wouldn't understand the answer, you know?"

"Everybody's dangerous. Life's dangerous," Quacks retorted.

"Not like women, pal. Women have lots of troubles. They *are* lots of troubles. Their purpose in life is to *stir up* trouble. See what I'm saying?"

"You had trouble with women, Mitch?" Buck Nelson asked.

"Everybody has. You got a woman, you got trouble."

"That's for damn sure," said Don Bell. "When I think of all the things women have done to me. Women," Bell snorted hotly. "And I don't want to talk about them, either."

"Deception," Mitch says. "What you see isn't what you get."

"You're just a poor judge of women, that's all," Quacks said.

"Let's get an example for you there, Quacko, so you don't stay in that cloud of confusion you're in. Let's say it's a nice day, just like this, and you're driving down a country road with your sweetness. A summer day, OK? You have the windows down, she's in a halter top and shorts, looking good, hair flying, hayfields, apple orchards lining the road, irrigation ditches full of clear water. Got the picture? Now. How many women do you know would turn to you and say, 'Hey, baby, stop the car, take me in your arms and carry me out into the grass, lay me down, and fuck me till I scream'?"

A stir rippled through the group as interest quickened to such a novel thought.

"There's probably some," Quacks said, in a voice short on conviction.

"None! That's how many," Mitch said. "This is more like how it would go: 'I heard they're expecting a big apple crop this year—juicy, big, luscious apples.'

"You keep drivin' on down the road and say, 'Uh-huh. Are the apples ripe already?' Silence. Next thing she turns and says, 'You never talk about the things I like to talk about.'

"And you say, 'What?'

"'You're never interested in what I'm interested in.'

"'I'm not?'

"'No! And don't act so surprised, either. You know what I'm talking about.' Then she folds her arms over her chest and looks the other way out the window.

"'Honey, I'm sorry,' you say. "'I guess I missed something.'

"'You're always *missing* something,' she says. 'Just forget it!'

"Next thing you know you're in an argument, and you feel bad, so you pull off the road, give her a hug, talk soft, decide on a walk in the orchard. A few minutes later she's flat on her back in the shade of an apple tree, and you're nuzzling on her nice like . . . That's deception!"

"Bullshit," Quacks cried. "That's not deception. They're just a lot more discreet than we are."

"It's because they feel so guilty," Troop added sadly. "Society did it to them."

Mitch came back. "They're warped, I tell you. You show me a

woman—any woman. I don't care how self-confident they come across when you first meet them. After two years you'll find out they're the pit of despair."

"That's a little hard on the fairer sex, isn't it?" asked Buck.

"I got a hard-on," Quacks chortled. "Had it for two weeks."

"Nice talk," Seiler said. "No wonder women are leery of smoke-jumpers."

"You're your own problem there, Quacko," Mitch went on. "You think women are something you can figure out. They're not! Women are insecure. They don't choose it, they just are. It's only a matter of time until they find the genes linked directly to all this hocus-pocus of human reproduction."

Lying there under the plane, drifting in reverie, I could see an open window, white birch trunks, a smoky sunrise. Twelve hours before, I had been with Sally making love under the big green quilt.

"You're fucking weird, Mitch," Quacks insisted. "Ha, ha, ha. That's insane."

Before me had been Sally's body, firm and petite, with clear, silk-smooth skin; her small breasts perfectly shaped, her nipples surrounded by delicate clusters of cocoa-colored freckles. Sally was rose petals, strawberries and honey, satin-tummied, and needed no pillow under her firm round buttocks.

"Look at history," Mitch continued. "Take the Vikings. They get hitched up, OK? The woman gets pregnant, right? The nagging gene kicks in and she starts in on the old man. It's a long winter up there on the coast of Norway, and by the time spring rolls around he's half goofy. So he jerks on his horned helmet, grabs a spear and shield, and takes off. Why? The wife wants more! More stuff. More stuff to make life better for her and the kids. OK, OK, he says, I'll get more."

Sally and I had snuggled there kissing. Cottonwood seeds floating on pine-scented air moved in and out through the window and landed softly about the room. After a particularly long kiss, Sally had taken my right hand and placed it firmly against her breast.

"The Vikings plundered half of Europe for two hundred years trying to get their wives to quit nagging. The ones that didn't get killed returned home with a boatload of booty for the family. The wives of those who died simply moved in as second, third, or fourth wife to the ones with all the goodies. Hey, no problem. That's history."

"That's absurd," cried Quacks. "You can't reduce the entire history of the human race to some cartoon notion of love and bonding."

Except for small whimpers, Sally lay quietly, her large eyes opening occasionally, blinking once or twice, then closing again.

"Read your science there, Quacko. You'll see. Evolution favored the aggressive, over-hormoned, not-too-bright male who was good with a spear and the female who constantly sensed the need for more security for her offspring. Aggressive males and insecure females eventually got the most stuff. They also had the kids that lived long enough to reproduce and project those same genes into the next generation."

In a comfortable confidence Sally and I had made love, then curled up in a damp clutch, her round bottom tucked tightly against my stomach.

"It was that way for thousands of years, maybe millions. It's in your loins there, Quacko, not your brains."

"I've never heard of anything so ridiculous," Quacks scoffed. "Another crackpot turning a little knowledge into a dangerous thing."

"It's the truth, lover boy. Don't advertise it, 'cause it sure as hell won't get you laid. Treat women with patience, take them shopping every chance you get, buy them everything you can afford, and realize that it'll never be enough, and that if it ever came down to it, she'd feed you to the wolves the minute she sensed she could do better with another, more aggressive man."

"You hate women," Quacks protested. "We're not cavemen anymore. This is the nineties! Women appreciate sensitivity."

"Horse shit!" Mitch insisted. "They give lip service to the sensitive-man trip, but inwardly they see them as weak pieces of shit."

"Lip service," Tyler Robinson moaned. "That's what I need."

Don Bell squirmed around, agitated. "Shit," he said with great finality. "Women! Is that all you can talk about? It was women that ruined my life. Do you have any idea how much money I've spent on women?"

"All you ever had," Mitch stated flatly, without turning to Don.

"All I ever had," Bell mumbled. "All I ever had. Yes, exactly! That one in Portland was the absolute worst! Drove me nuts! Drinking, smoking, and all the time flirting with nerds. Insisted on having *Playgirl* magazines *right* in the bed with us. Having friends over that were the most ridiculous worms you ever saw. Men friends! Always coming around talking and acting nice. I wound up throwing one of them out her front-room window—chicken shit city nerd. *Grrr... women.*"

Bell's storminess brought on a brief silence.

"What do you think'll be in those sack lunches?" Kubichek wanted to know.

"Women! Women! That's all I hear," Bell suddenly blurted. "All my life, they've been nothing but trouble! Caused me to go out on rainy nights, looking for them in the dark while they hid in the bushes not wanting to be found. Called me a racist and a redneck. Caused me to start fights in restaurants, chase flirtatious nerds down the streets, run from the cops, and constantly always wanting to eat out all the time. Chinese, Italian, Mexican, Mongolian, eat out, eat out, all the time, eat out. Do you have any idea how much money I've spent on women?"

"All you ever had," Mitch said again.

"That's right," Bell said, settling into a bewildered calm. "How'd you know?"

Just then, a Fairchild F-27—a twin-turbine, medium-range passenger plane—roared over the airstrip, banked out over the river bluffs, turned on final, and landed. It came screaming down the runway trailing a cloud of dust and blew past us, spraying gravel and leaving us coughing and cussing.

"How in the... who the... *Shit!*" Bell screamed. "Probably a damned *woman* pilot."

The plane continued to the end of the runway, turned around, came back, and parked right across from us.

"Boy, that'd make one hell of a jump ship," Buck said. "You could put the whole crew in it."

Slowly and silently a hydraulic stairway arced down from the top of the fuselage and came to rest a couple feet from the ground. To our utter amazement, a uniformed stewardess came down the stairs carrying a little stool and placed it at the bottom.

"Shit, they must have been hijacked to wind up in a place like this," Mitch said.

Down the stairway they came, as if the F-27 was a charter bus— America's grandmas and grandpas come to see the last frontier. They wore slacks, sport clothes, and carried cameras of all sorts. Soon a number of them trained their eyes on our little group lying in the gravel underneath *Jump 17*.

"Tourists!" said Quacks. "Can you fucking believe it? Tourists in Eagle, Alaska."

Bell snorted, "Wonder what they'll do. Take a side trip to Chicken?"

A gray-haired man in Bermuda shorts strode over and told us he was a retired wheat farmer from the Midwest. It was a tour group out of Anchorage, he said. Airplanes were better than buses because they were faster and you could get more for your money flying around to see the country.

"Let me give you fellas some advice," he said, looking down to consult his puffy fat hands. "You ought to consider taking such a vacation," he went on. "See Alaska from the air. You can't see much from the ground. Alaska's a great place to fly, beautiful mountains, and country you wouldn't believe."

"Yeah, well, we get in our share of flying," Buck said. "We just flew out from . . ."

"Tonight," he interrupted, "we'll be staying in Fort Yukon. Then on to Fairbanks, Prudhoe Bay, and Nome—all that country in just four days."

We told him that was nice and wished him happy travels. He strolled away, exceedingly pleased to have impressed us with his plans.

"Unbelievable," Mitch muttered. "*See the country from the air.*

Shitfire. What the fuck does he think? That we drive this thing on the roads?"

"Tourists," Bell hissed. "This tops it all. Tourists telling *us* how to see the country. Who does that fat-assed, frog-eyed midwestern wheat nerd think he is, anyway? He doesn't know shit about Alaska. So he'd better stick to his little square fields, flat roads, and his nerdy wheat-farming idiot neighbors . . . *Grrrr.*"

Another dust cloud appeared. Not on the airstrip but on the road from town. Around the corner of a willow patch came a school bus, rattling its way toward us, pulling to a stop a respectful distance from the big plane. Its motor went silent. The door folded in on itself, and the driver stepped down.

"Praise Big Ernie, would you look at that," Quacks said. "Now how could anyone say anything harsh about something like her?"

"Easy there, Romeo," Mitch said. "She's the kind that would make quick hash of you."

"Damn," Bell said. "She looks like an angel."

"I'd tunnel all the way underneath this runway for a peek under that dress," Quacks said.

This bus driver wore a floppy straw sunbonnet. Her dress was of some sheer, airy fabric, V-necked and sleeveless, and covered with pastel pink and lavender rose clusters on a light blue surface. Her arms were tanned, like her legs, and her black hair hung straight to her waist. If it hadn't been for the hair, I would have sworn it was Sally. The jumpers—all but Mitch—were instantly bewitched. If the woman had arrived driving a team of six white horses, she would have made no greater impact.

"It's like a dream," Troop said. "Like we've died and gone to heaven."

A light breeze whipped up a little dust and pressed the dress against the driver's body. She turned her back to the wind, to us, and the dress swept up between her legs.

"Grrrrr . . ." said Bell. "Let's get the hell outta here."

Down the road came the BLM pickup. Everyone, not just Ku-bichek, knew what that meant. Lunchtime.

We got back under the wing of the plane and dug into our lunch sacks as the tour group began loading into the bus. The lunches smelled good: three bologna-and-cheese sandwiches, an apple, a can of Pepsi, a can of apple juice, a bag of potato chips, and two candy bars. I had my first sandwich nearly eaten and about half the Pepsi drunk when the radio squelched and began talking.

"Jumpers—Eagle dispatch."

"Jumpers," Gould answered.

"Roll toward Canada."

Minutes later, we were off the ground, having left behind a few sandwich wrappers and a dust cloud filled with coughing tourists.

As we lifted out across the Yukon, John Gould turned from the cockpit and yelled. "We're just fifteen out—go ahead and do your buddy checks."

After our checks were completed, I reached into my leg pocket and retrieved what was left of my lunch. The remaining sandwiches had been reduced to damp lumps of bread and meat. The chips were pul-verized, and I'd poured the remains of my Pepsi onto the runway. The juice was still OK, so I drank that and wadded what was left of the sandwiches and chips into a ball and ate that, too. A note was passed back to Troop.

Going to original fire B-405—12 acres, heavy black spruce—
100% active—rolling. Air Attack and air tankers en route

Pulling up over the fire, we were instructed by Air Attack to hold in an orbiting pattern for a few minutes while the tankers finished their initial drops. From our position we could see both our fire and the one Dalan Romero was on. The smoke from Dalan's fire lay flat and hazy, indicating they had it contained for the time being.

Ours, however, had another towering smoke column and was

rolling south on a high, dry ridge covered with scattered black spruce and caribou moss. A river ran perpendicular to the path of the fire about three miles farther south, but it appeared unlikely that the river would be able to stop the fast-moving head. There were no other natural barriers in any direction. Worse, I couldn't see one single water source within workable distance. We had only twenty-six hundred feet of hose. The only thing in our favor was the location of the helispot the first two jumpers had built before being run off by the bear. The helispot had not yet been burned over, so we planned to use it as our jump spot. That would prove handy in securing our jump gear and likely prevent some damage to parachutes as well, since we would, we hoped, be landing in the opening and not in trees.

John Gould came crawling back over the jumpers, ready to go to work.

"OK, you guys," he yelled. "Take a good look at this. You're probably gonna have your hands full. I don't see any water anywhere."

John went to the door. "Guard your reserves!"

The door swung open. Up close, the smoke column boiled rapidly skyward. *Jump 17* bounced roughly in the turbulence.

The first set of streamers sucked right into the hottest part of the fire. The second set landed near the helispot; a third set did the same. John raised two fingers, indicating to the first two jumpers that it was time to jump. The plane suddenly hit bad air, rocked left, dropped a wing, went weightless, then slammed us down hard. Troop and Clouser moved cautiously to the door, bracing themselves so as not to fall out.

We pulled around on final, with our lineup just off the right flank. Near the tail, backing flames eight to ten feet high were eating their way steadily toward the jump spot. The fire behavior was quickly becoming radical with ten- to fifteen-foot flame lengths along the flanks, and thirty- to fifty-footers at the head. The main ridge ran due north and south. The helispot—the jump spot in this case—was at the north end, approximately one hundred yards from the fire's edge. A bit close for comfort, but the wind was steady out of the north.

John pulled his head back out of the slipstream and sent Troop and Clouser out the door. Hurriedly, John retrieved the static lines for the two drogue chutes, unhooked them, threw them in the back, then turned to Kubi and me.

"Are you ready?"

"We're ready!"

"Get in the door," John said, patting the threshold with the flat of his hand. Again, his head was out watching Troop and Clouser to see if any corrections were needed concerning lineup, exit point, and wind drift. *Jump 17* circled the smoke column, banking right in a slow arc. As we came around, Troop and Clouser disappeared behind the smoke. I did my four-point check, making sure my rip cord was in plain sight.

"OK, Murry," John yelled. "Looks like about three hundred yards, mostly down low. Stay wide of the fire. You should be able to pick your approach quartering in. If you overshoot, that's better than landing too close to the fire. Any questions?"

I shook my head no. John looked around at Kubi, who stood behind me, his right hand on the door above mine.

The plane leveled out on final.

"Get ready!"

I leaned back, bracing myself with both hands, eyes fixed on a point off the wingtip somewhere in the smoke. In one corner of my eye, I could see a mountain of fire raging, in the other was John's hand poised behind me. Suddenly I was falling, looking for my rip cord then back at the plane, then down at the fire. I pulled. My chute opened, and I turned to watch Kubi's do the same.

Far below, my eyes found Troop and Clouser on base leg approaches with nothing but fire below them. Their progress against the wind was steady as they cleared the edge of the fire and moved over thick spruce toward the jump spot. With the wind increasing as I descended, I set up on a shorter final approach than they had. They landed near the center of the seventy-five-foot-diameter helispot. I pulled around on final at about five hundred feet. At 250 feet, the wind gusted due to a large flare-up off to my right near the tail. I let up on the toggles and the chute recovered into the full-run configuration for maximum penetration against wind. In the next instant I hit some down air and began sinking. Reefing down on both toggles, I held them stiff-armed as low as I could until the chute flared into near-level flight just in time to clear the tops of some forty-foot black spruce at the very edge of the spot.

In the spot's center was a space about twenty feet in diameter where all the debris had been removed. The remaining area contained a horrible jumble of jackstrawed trees lying every which way, two to three feet deep. Under them were the sharp stobs of hundreds of cut-down trees. I was headed for a point about fifteen feet short of the center clearing when I heard crashing and yelling behind me. I had problems of my own. I was going to make the helispot but not the center clearing. A spruce log lay crossways in my path, waiting to whack me somewhere just above the knees. If I lifted my legs, I ran the risk of busting my ass. If I didn't, I could break a leg. It all happened so fast I don't know whether I made a decision or simply reacted, but when my feet hit the ground, I dove headlong over the log, leaving my un-weighted legs to impact the log with minimum resistance. My plastic knee guards took most of the shock, sending me flying end over end. When the crashing finally ended, I was wedged head down under a big spruce, tangled up in parachute lines.

"Nice flip," Troop said brightly. "You OK?"

"I doubt it," I said. "Wait'll I stand up. Then I'll tell you."

"Fuck this!" Kubichek yelled from somewhere off in the woods. "Fuck this bullshit!" he huffed, grumbling about downwind, ground wind, and such.

The rest of the jumpers came in. Quacks and Robinson landed out by Kubi. Bell, Mitch, and Buck made the center clearing. Seiler took a dive in the jackstrawed slash like me. Shortly after I removed my gear, Troop started talking.

"As soon as you're ready, I want you to take charge of the left flank," he told me, taking a bite of a shiny red apple he'd salvaged from his lunch. "Once we get the cargo, I'll send three guys your way, but in the meantime I want you to check the head and scout that drainage to the east for water. Let me know as soon as you find some."

Troop took another bite, lifted his radio to his mouth as if he were about to say something to the jump ship but then changed his mind. "Once we get the tail secured, I'll bring the rest up the right flank."

Just then, *Jump 17* roared over at two hundred feet, and we watched Gould kick out our first fire pack and a five-gallon cubitainer of water. Slowly the cargo fell from the door, then accelerated, streaming toward us. A hundred feet off the ground, the chute opened and snapped it up under canopy.

"Good shot, *17*," Troop radioed. "Keep 'er comin'."

I arranged my PG bag for fast travel, stuffing in what was left of my lunch and a few other goodies I'd ratted from the fire pack. In five minutes I was heading up the left flank.

The fire was burning very hot, so I kept out between forty to fifty feet—far enough to remain comfortable, yet close enough to keep my eye on it in case conditions suddenly worsened. It would only be a short time until the tail of the fire would burn through the retardant that had been dropped. We needed people on the line quick. The drop up the left flank extended about two hundred yards and then thinned out—another problem. We needed people there, too, or the fire might flank out around the end of the retardant line and eat its way back toward the jump spot.

Farther along the left flank, the flames were higher and progressively more erratic. About two hundred yards ahead I could hear the head of the fire sizzling, snapping, and roaring. Then—a warning. The roar was rising, the caribou moss crunched loudly underfoot, the spruce smelled strongly of resin, and embers fell out into the unburned woods trailing thin plumes of smoke. The air was beginning to tremble. Assessing a fire situation requires an abrupt assimilation of senses, the quick collaboration of experience, intuition, and perception. It's never just one or two factors—mostly it comes down to a gut feeling, and my gut feeling was that the head of the fire was developing into a possible blowup.

I called Troop and reported what I was seeing on the left flank.

"Help's on the way. Found any water?" Troop asked, out of breath.

"Haven't had time to look . . . but I doubt we will."

"If the winds keep shifting, we're not gonna be able to protect our gear," Troop said.

"From what I'm seeing up here," I said, "unless we get several loads of retardant, we haven't got a chance, wind shift or not."

"*Air Attack 363*, this is Troop."

"Troop, *363*."

"How far out's our next load of mud?"

"Standby one."

"I hope it's soon or we're gonna lose the helispot and our gear."

Air Attack came back. "Your only safety zone is into the fire, I suppose you know that."

"Copy," Troop gasped.

I continued toward the head, dodging around spot fires. If the blowup came, my escape route would be back up the left flank and into a burned area cool enough to survive—although I hadn't really seen any like that since leaving the tail. Attempting to escape into the unburned woods could prove fatal. Once a fire begins to run, the safest place is usually somewhere inside it.

Some of the spot fires were bucket-sized, others were the size of pool tables. A couple were already volleyball courts. All were burning hot. We were at that critical stage in fighting a fire, that time when we needed to decide if we had a realistic chance of catching it without taking too much risk. Catching it depended on several factors, but two were paramount. First, the head was still small enough that it could be temporarily knocked down with two or three loads of retardant. Second, we needed more jumpers.

Troop was busy on the radio, huffing and puffing around the tail.

Air Attack leading an air tanker. *Photo by Mike McMillan, © Spotfire Images*

A bird's-eye view of the mighty Yukon River. *Photo by Davis Perkins*

Tanker 84 working a fire in Alaska. The drop did not miss its target. *Photo by Dennis R. Terry*

A burnout at night on an Idaho fire. *Courtesy of the Bureau of Land Management*

After hanging up in a tree, a jumper descends on his letdown rope. *Courtesy of the Bureau of Land Management*

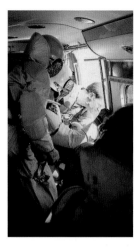

The spotter looking out the door. *Photo by Davis Perkins*

The jumper's view. *Courtesy of the Bureau of Land Management*

A jumper under canopy with killed drogue trailing behind. *Photo by Craig Irvine*

Extracting jumpers and their gear from an unsafe jumpspot.
Photo by Mike McMillan, © Spotfire Images

A blowup in a stand of ponderosa pine. *Photo by Mike McMillan, © Spotfire Images*
INSETS (TOP LEFT) A typical Alaskan gobbler. *Photo by Davis Perkins*
(TOP RIGHT) Snags left standing at dawn. *Photo by Davis Perkins*
(BOTTOM) The crash of *Tanker 138*. The pilot walked away unharmed. *Courtesy of Davis Perkins*

INSETS (TOP) Murry Taylor on the third day of the Clear fire. *Photo by Chris Woods*
(BOTTOM) In Alaska, a jumper beats out a fire perimeter with a spruce bough. *Photo by Steve Nemore*

The old-timers in front of the standby shack. From left to right: Erik the Blak, Don Bell, Jay Peterson, Pat Shearer, Murry Taylor, and Rod Dow. *Courtesy of Murry Taylor*

Murry Taylor on his 200th fire jump, five miles from the village of Rampart, July 1999.
Photo by Tony Pastro

Everyone was with him trying to protect our gear. I pushed on around the head. My safety zone was now back down the fire line at least three hundred yards—too far. There had to be something promising somewhere, something to hang our hopes on.

Rounding the head, I became temporarily confused by the lay of the land. The head appeared to be slowing down in an area of about two acres, a broad saddle, beyond which the ridge dropped away. At times the smoke was so thick I couldn't see more than thirty to forty feet. I began to run. The flames grew louder. I couldn't hear the radio. I was caught in a fifty-yard strip between the main fire and a large spot fire located on the ridge crest south of the saddle. Both fires were dropping embers all around, starting spot fires.

Blowups may result from a number of factors, not the least of which is wind. Winds are often associated with approaching weather fronts, topographic features such as canyons and river channels, the proximity of cumulus clouds, and, in Alaska, glaciers and ice fields. These are all factors apart from the fire itself. Fires also blow up without any apparent external reason if conditions become extreme enough within the fire itself. Other factors aside, blowups from fires burning on high ridges are less common than those from fires that start in canyon bottoms with long upslope runs. This problem is compounded when a fire in a narrow canyon spots across to the opposite slope, risking mass ignition. In the most extreme blowups, wildfires can consume thousands of acres in just a few minutes.

I hightailed it out of there just before a frightful ball of red fire and black smoke swelled overhead behind me. Embers came streaming down on my shoulders, burning my neck and the back of my fire shirt. I circled west in front of the head again, still trying to see if there was a way to contain the whole thing. Beyond the saddle was a rocky point, where the ridge began its drop downhill to the river about two miles distant.

"Taylor—*Jump 17*."

"Taylor, go."

"OK, Murry, we're coming around. Are you at the end of the ridge there, where she's really cooking up?"

"That's affirm! Southwest corner."

Digging my army signal mirror out of the top flap of my PG bag, I flashed 17 just as they winged through a patch of blue sky framed between smoke columns.

"John," I yelled into my radio just as a bunch of spruce torched off up the hill. "We need water, but it's got to be fairly close."

"Roger, but you're not going to find any. We've been looking already. Your closest water is the river."

I asked for a size-up of the head. John confirmed what I was seeing on the ground, relayed some of it to Troop, then departed for Fairbanks. The only plane left in the area was *Air Attack 363*—Rob Collins.

Troop called. He told me that the tail was looking better and that there were three Canadian air tankers en route from Whitehorse, Yukon Territory.

"When they get here," Troop said, "work them all up your way. Also, we got ten more jumpers twenty minutes out."

I continued around the head until I was on the right flank. Conditions remained marginal, but I felt in less danger. The dip in the ridgeline had stalled the head's forward advance. For the time being, the fire was confined to the top of the ridge and would have to burn downhill to spread farther. Moreover, off the top the winds were calmer.

Following the intense activity in the saddle, the head of the fire began sucking back into its middle. Visibility around the flanks improved. The big spot and the main fire were now one. I circled the area snuffing spots with my feet and Pulaski. Ridding the area of spots would make it safer for the other jumpers to move in without having to worry about spots below them.

I worked my way back to the rocky point where I could see more country. Broadening to nearly a quarter mile, the ridge became almost flat in places and was covered with yellow caribou moss and a few evenly spaced young black spruce. The country lay in the golden light of midafternoon, empty and pristine. Across the river, pale blue shadows rimmed the edges of a broad yellow mesa that stretched south under a big sky.

A scratchy voice came over my radio.

"Aahhgg," it began. "Fire B-405, this is *Jump 07* on Turquoise. You copy?"

It was Erik the Blak with our second load of jumpers.

They flew low over the end of the ridge, pulled into a steep climb, and began circling. I watched them jump, seeing them now and then through windows in the smoke, one or two parachutes gliding down, red-orange and white, vivid against the blue sky. In twenty minutes, the load aboard *Jump 07* was on the ground, all its cargo had been dropped, and the Blak was barking good-bye. As soon as *Jump 07* had cleared the area, *Air Attack 363* was back overhead with a message for Troop—the first Canadian air tanker was only five minutes out.

"OK, *363*, I copy. When it gets here, go ahead and contact Taylor. He'll take most of it."

The fire was snorting behind me, hissing and popping confidently and raising general hell, but it wasn't long until I heard the approaching drone of a large aircraft.

"Taylor, this is *363*."

"Go ahead."

"OK, Murry, we've got an air tanker—the first of three. Troop says they're yours. How on that?"

"Rob, we had a big spot here a half hour ago. It's apparently burned back into the main fire."

"Yeah. We'll drop down and give you a closer look."

"The best approach," I went on, "would be to go ahead and extend the retardant line down the right flank, being sure to save the last load to make a good V around the point at the end of the ridge. That way we can contain a lot of spots."

"Copy that. Let me get with this first tanker, and then I'll get back to you."

Air Attack made a low pass right over the end of the ridge, pulled up into a steep climb banking right, and disappeared beyond the smoke.

"Taylor, *363*. I'm thinking a couple drops on the head to slow it down first, then some flanking and then finish the head last. How's that sound?"

"Sounds fine. I'm out of the way."

"Any other people in the area?"

"Some jumpers will be coming up the left flank anytime now, but . . ."

"Leathersack, this is Quacks."

"Go."

"I've got five, and we're about halfway to you. The edge is beating down pretty well, but it'll be a while till we're up there."

"*363*, did you copy Quacks?"

"Yes, we did!"

"OK, then. Let's have some mud."

"Here we come turning final. This one'll be live."

"Copy. The line's clear."

The deep resonating throb of the approaching tanker always gives me a thrill, and this time was no different, putting the hair up on the back of my neck. From where I hid in the trees I couldn't see it; somewhere out there in the sky, an aerial fire bomber was dropping a wing and turning final, its speed increasing as it dived, the roar building, steadily growing louder and louder. In a flash of light the lead plane zoomed past, followed by the approaching whine of giant radials. Through the trees I saw a great form crowding the space between itself and the ridge, and then in a sudden amplified crescendo, the sky thundered, and the world around me exploded into a roar. Shuddering with the fiery exhaust of thousands of horsepower, the Canadaire opened her gates, sending a streaking reddish pink blur earthward. The roar moved off in the direction of the flight path, while overhead pink rain descended. I ducked low under a spruce and closed my eyes. When I opened them, the world and a good portion of my body was coated with slimy red snot. I hadn't told the whole truth about being out of the way but had chosen to endure the discomfort of being pasted with retardant in exchange for the benefit of being able to see something wonderful up close.

As *Air Attack 363* worked the head, Quacks and the five other rookie and snookie types from the second load came up the line full of themselves, laughing, carrying on, and eager to tell Old Leathersack all about it.

Yes! they told me, it was true. Things had been going really great for them. And, yes! It was also true that this smokejumping was, indeed, the greatest job in the world, and of course, wasn't it amazing how most people would pay a fortune to come on a trip like this? And furthermore, this is the way my jump went, and further still, wasn't it wonderful to be on a fire, kicking butt, making overtime, and were we not, in fact, going to take all our money during the coming winter and travel to all the great beaches of the world where we would gaze at the unbelievably beautiful, topless European babes?

I growled at them like an old dog at pups. "Quiet down a minute," I interrupted. "That's all fine, but we can't stand around here squawking like a bunch of ravens. We've got a damned fire to catch."

The radio squelched loudly, then started in.

"Taylor, *363*."

"Go ahead."

"Murry, that's it for the Canadian ships. I just called Eagle and they have *Tanker 84* off Tanacross a few minutes ago. He was assigned a 'load and return' to your fire. Copy?"

"OK, Rob, sounds good."

"We're low on fuel and have to head out. Anything else we can do?"

"Pass our thanks along to the Canadians for the impressive air show. Damn nice job."

"Will do. *363* clear."

I explained to Quacks and the new arrivals what I'd seen on my recon. In typical jumper fashion, each of them came up with an idea of their own. One thing you're never short of with smokejumpers is opinion. Two jumpers—two plans. Six jumpers—six plans. Thirty jumpers—thirty plans. My rookies and snookies spoke loudly at each other, waving their arms in the air, nodding in mock agreement, and listening little to what anyone else had to say.

I smiled inwardly at their exuberance and, in that moment, could not help but love them very much. They were fully alive and having fun. In the glory of their youth, their happy, young faces dripped with sweat, while behind their eyes they were still seeing grand visions of themselves, godlike, flying through the sky, tumbling to the ground, making plans against enemies, and going for a hike in the woods with their pals.

It was impossible not to notice that I wasn't quite like them anymore. I was still strong and skilled maybe but no longer powered by that same enviable enthusiasm. During the last few years I'd had to accept that I no longer occupied a position near the social center of our experience. I'd been edged to the outer circle, the old warhorse turned old goat—*Old* Leathersack. No longer one of the rising stars, I was now the ancient one, the wise elder, the voice of experience. But the pups aren't in the center either, really—that will take a few years until they more fully prove themselves. In the meantime they're moving in the right direction.

After my new arrivals had gotten most of the testosterone-adrenal-voodoo out of their systems, I directed them to our next course of action. Up the hill each one tore, diving into the work in a manner that not only suggested that it was *his* plan that we'd adopted, but apparently elated by the prospect that it was certainly one of the more brilliant ever conceived.

An hour after the Canadians had departed, *Tanker 84* arrived to reinforce the scratch lines we'd put in around the hottest part of the head and portions of the line along the tail. Busy cutting saw line, widening scratch lines into finished hand line, falling snags, and scouting for spot fires, we worked the hours away while the smoke slowly thinned. Troop was working Mitch, Bell, Buck, Seiler, and the rest of his people at the tail and up the right flank. From the sound of the radio chatter, things had apparently turned the corner in our favor.

After a while Troop called and asked if we were in a position to send four people back their way. We were and I did.

By then it was seven o'clock with the sun swinging low into the northwest. Our monstrous column of smoke had drifted off lazily to the southeast over the river country and the high mesa. The perimeter of the fire was a crooked mess of skinny, hot fingers, burning clusters of logs, hot ash pits, and smoke. Tired and sweaty, we worked on until nine. Then Troop radioed again. His voice was flat and spent, but as always infused with humanity.

"Hey, Murr, ol' buddy. How's the dragon lookin'?"

"Hey, Troop, I think she heard you were fire boss and just decided to give us a few hot blasts and roll over and die . . ."

"Naw, it's you guys she heard about. Things are lookin' up all right. We got another problem now, though. A few minutes ago a black bear showed up and chased Kubi away from a fat boy box. Now he's drug it off somewhere."

"Probably the same one that got after those other jumpers," I said.

"Yeah, probably. Tell you what—I'll send Buck and Seiler down that way with some rats and water. It's time we had a break. We got two crews coming. Should be here by noon tomorrow. Let's figure on a full night securing the line. I don't see us getting much sleep, so let's eat now and pace ourselves for an all-nighter. How on that?"

"You're the man, Troop."

"All right then, let's do it that way. Troop, clear."

Patrolling back and forth, we spread out and worked the hot spots closest to the perimeter first. Now and then another finger would flare up and we'd hustle over and beat it down with spruce boughs and dig more hand line. The heat of the day had passed. The fire cooled, and the ridge reclaimed some of its calm. Forty-five minutes later, the guys with the food showed up, and we hooted to each other our signal to gather up.

The rock outcropping at the end of the ridge made a perfect spot for a rough fire pit, so we got busy collecting firewood and dragging up logs for seats. Low and brassy, the sun appeared below some distant clouds, turning the birch trunks stark white, while showcasing the bluffs along the river in soft yellows and greens.

I ate two instant soups, a packet of Ry-Krisp with peanut butter and jam, and three large, rather burned chunks of flat nose—canned ham to others. I finished up with half a giant Snickers bar and another cup of coffee.

During dinner I noticed Al sitting off to one side of the group, holding his signal mirror in one hand while pulling down the skin under his left eye with the other. Buck Nelson went over to him and bent down to take a look.

"Damn it, man! What is that—a canoe paddle?"

I got up from the fire and went over to look myself. A half-inch sliver was lodged deep in the cornea. Buck worked for a while with a clean piece of gauze, pouring water in then dabbing at it with the corner of the cloth. Al looked miserable.

I suggested privately that maybe we could contact Eagle and get him out on a helicopter. Growling low like a badger, he shot me a white-hot look for even thinking such a thing. When I further suggested he consider resting by the fire and keeping the coffee hot, he jumped up, ornery-like, and took off for the fire line to begin digging trenches down to mineral soil. Apparently the sliver had shot out from the chain saw while Al had been cutting line earlier in the day. Later on, I asked him again about taking it easy, but he just hissed and walked away.

One night on a stretch of fire line overlooking a big gray valley, Al had once told me something.

"My dad used to get drunk when I was a kid," he'd said. "He'd take me out in the backyard, jerk me by the arm, and point up in the apple tree and say, 'See the angels? See the angels?' He'd twist my arm, pull me close, and say it rough. 'See the angels, there on the limb?' And I'd say, 'Yes, I see them.' Then he'd let go. One time he got so messed up he held a pistol to my head and pulled the trigger real slow, then laughed when it'd click. 'You got to learn how to be a man,' he said. 'You got to learn how to be a man and not be afraid.'"

Twenty-five years later, Al was out here in the woods working his ass off—one of the best smokejumpers that ever lived. In the meantime his father had died and gone off somewhere with his angel friends and would never know shit about how great a boy he had.

On into the night we patrolled and improved the line. The sun fell below the horizon. The sky went pink. I don't know how the rest felt, but fatigue and a full belly were catching up with me. Maybe it was the quiet of that country, beautiful and empty, I don't know. Whatever it was, it brought with it an unmistakable loneliness.

I began digging in a hot ash pit, cutting and hacking with my Pulaski at a tangled mass of smoldering roots as big as three pool tables. Roots, one to four inches thick, ran every which way. The task suddenly loomed before me as so overwhelming, I wished I hadn't found the thing at all. But it was close to the line, and if I didn't dig it out, one of the other bros would have to. The work was god-awful tedious, hot, and smoky, and with the prospect of a sleepless night ahead, the entire situation struck me as absurdly dismal. After having spent twelve hours with a torrent of adrenaline pumping through my veins, the comedown had left me feeling awful.

Shit, I thought. *The hell with this. Way out here in the middle of nowhere, beat to death, tired, wild-pig dirty, nothing to look forward to except ash pits and smoke, getting your face scratched, maybe your eyes put out. Eating burnt flat nose, drinking bad coffee, shitting in the woods, wiping your ass with sticks and dry moss, watching over your shoulder for bears, and having to stay up all night by yourself with no one to talk to. What a weird fucking way to live. And for what? Another long day tomorrow and the same old shit all over again.*

A twig snapped and I whirled around, thinking it was that damned chowhound bear. It wasn't.

"Hey, Troop. My hero."

"Hey, Murr. How's it goin'?"

"Oh . . . I was just thinking what a fucked-up job this is," I said, continuing to flail away at the root wad.

"Oh yeah? Don't you like this?" Troop asked, dismayed.

"I like women, hot tubs, and clean beds, not an all-night assignment stumbling around in sixty acres of smoke and hot ashes."

"Hey," Troop said, smiling big. "Good job down this way. We kicked her ass, huh?"

I looked at Troop, his fire shirt, PG bag, and hard hat all caked with dried fire retardant, his face black and greasy. Red-eyed, black-lipped, and white-toothed, his shoulders slumped heavily. Gray hair stuck out from under his hard hat.

I told him, "You look like hammered shit."

"What else is new?" he asked. We both laughed.

"Good job to you too, buddy," I finally said after we'd calmed down.

"What're we gonna do about this bear?"

"With crews coming in, you better order a shotgun and slugs."

"Yeah, I did that. Maybe he got enough to eat out of that fat boy."

"There's no such thing as bears that get enough to eat," I said. "He knows what's in those boxes, and he's made Kubi run. He'll be back."

Troop looked off down the fire line, then turned and looked back the way he'd come. After a while he sat down, and I went back to digging.

"Hey, Troop, what's on your mind?"

Troop took out his handkerchief and blew his nose loudly. "I was just thinking about Mouse."

I stopped digging and looked over at him. Troop's usual brightness was gone. He just sat there staring at the ground. For all his cheer and indomitable spirit, I'd seen Troop's shadow side before.

I stepped clear of the ash pit, walked over to my PG bag, pulled out my canteen, took a drink, and thought about Mouse.

In an elegant stand of birch near the top of Birch Hill, where we do most of our practice jumps, a plaque is nailed to an old tree. Troop had made it out of plywood, cutting scalloped curves along its borders, burning it with a blowtorch, then smoothing the rough edges with sandpaper. Mounted in the upper left corner are two of Mouse's U.S. Marine Corps Eagle, Globe, and Anchor Pins. In the upper right corner are his Alaska and Cave Junction smokejumper belt buckles. Along the right side is the rip cord Mouse pulled on his last jump; within the handle is glued a pin commemorating his 250th jump. On a small shelf sit Mouse's smokejumper-model White boots, the tops of which have been ravaged by red squirrels needing something salty in their diets. Near the center is a framed photo, the image of which has faded, leaving behind only a glass-covered piece of white paper. All these items are glued to the plaque and encircle the following words:

MOUSE WAS OUR CHEERFUL FRIEND WHO BECAME A SMOKEJUMPING LEGEND. HE WAS A CAREFUL AND SKILL-FUL JUMPER BUT ALSO TOUGH, LOUD, AND HUMOROUS. HE WAS THE SMALLEST EVER TO WEAR THE EAGLE, GLOBE, AND ANCHOR.

ONLY 4′ 10″ TALL, HE BROUGHT HIS HUMANITY TO THE BATTLEFIELDS OF VIETNAM FOR THREE YEARS. IN THAT LAND HE WAS KNOWN AS LITTLE BIG MOUSE HON-ORABLE SIR. IN 258 SMOKEJUMPS HE COVERED 15 STATES FROM ALASKA TO GEORGIA. HE WAS KILLED SEPT 6, 1985, WHILE SKYDIVING AT NORTH POLE, ALASKA.

MOUSE LIVED FULL AND FAST. HE WAS KILLED LIVING LIFE HIS WAY.

"You weren't here when he died," Troop said, "but it hit everybody hard. Bell had been his roommate all that summer in the barracks. Don went off into the woods and didn't come back for two weeks. In the end, all we had was a little jar of ashes no bigger than a can of pork and beans. We mailed them back to Cave Junction, and then in the fall we had a memorial ceremony at the old jump base. After people talked about Mouse, a fake fire call came in requesting one man to the Kalmiopsis Wilderness Area. We 'borrowed' a Forest Service plane and made an outlaw flight over the Green Wall along the Illinois River. When we opened the window and held the jar out, Mouse streamed out in a long line like a shooting star."

Mouse had been killed in a freak skydiving accident. He had just opened his chute when another skydiver collided into him at 120 miles per hour. Both died instantly. I wasn't smokejumping during the summer of 1985, but I knew Mouse from his Cave Junction days and had jumped fires with him.

Allen "Mouse" Owen had petitioned the U.S. Marine Corps and several congressmen to allow him to enlist in the Marines, even though he was too small to meet their standards. Finally they allowed him in. Mouse went to Vietnam, fought there, and returned a decorated soldier. His picture made the cover of *Life* magazine, along with a feature article on "The Littlest Marine." Later he petitioned in the same fashion to become a smokejumper.

Troop took Mouse's death the hardest. The marine thing must have figured into it. After returning home from Vietnam, Troop traveled from one end of the country to the other paying personal visits to the fathers, mothers, and families of those killed under his command. He used his own money and leave time. As a second lieutenant and the commanding officer of their dead sons, he felt it his responsibility. He told them stories about their sons in battle and how they had been brave and fine men until the very end. Some of the families were grateful and Troop spent days visiting with them. Others, like one elderly widow living way out in the Kansas plains, left him standing at the screen door knocking while she rocked in her chair, repeatedly ordering him to go away.

"Hey, Trooper," I said, in a stern attempt to break the funk he was in.

"I don't know much about death, but I do know this. Mouse thought the world of you, and the way I figure it, your men in Vietnam did, too. Mouse spent a lot of time close to death. He also understood that our time here is precious, and he made good use of all he had. The memories we get to keep, Troop. The rest we have to let go."

I went back to digging smoldering roots while Troop just sat there quietly. After a moment, he got up and started off down the fire line.

"Troop," I said, turning to watch him go. "You're a damn fine man." Not looking back, he spoke something in the direction he was heading. As near as I could tell he said, "I got memories enough. I just wish I could forget."

13

July 3   Yukon Territory, Canada

At 6:45 I woke from a deep sleep. I lay there feeling rugged, unciv-ilized, complete. We had worked until 2:00 A.M., yet surprisingly, I felt rested. I heard a noise and looked up to see Troop walking the fire line. Sounds of digging came from down the line in Troop's direction, so I got up, folded my sleeping tarp, repacked my PG bag, and headed for the rock outcropping.

The sun burned in a clear eastern sky, the air was warm and sweet. A solemn calm lay in the land; drift smoke settled in the canyon. At camp I found the first risers milling around a crackling fire, heating cans of water, making coffee and instant cereal, and grumbling about their burned chunks of canned ham.

"Shit!" Mitch said, holding up a smoking black object, then taking a bite.

Kubichek wanted to know how soon we could get some fresh food.

"Oh, shut up and quit your whining," Mitch admonished. "You just had a sack lunch yesterday. This ain't Outward Bound, you know."

Al sat off by himself—his eye red and swollen—not saying any-thing, just looking out down the ridge.

"Man, I gotta get laid when I get back to Fairbanks," Quacks said.

Mitch took another bite of his flat nose, then glanced hopelessly at

Quacks. "Watch out there, Romeo, or you'll have another one kanga-rooin' up and down on top of you before you know it. Your IQ will fast plunge into the *ohhh, I'm in love* range, and the next thing you know you'll be seeing your little sweetness heading out the door with your checkbook and credit cards."

Bell shifted around the campfire trying to avoid the smoke, cussing to himself as he went.

"Next thing you know there, Romeo," Mitch chided, "you'll be talking marriage."

"Marriage," Kubi choked. "Why would anyone do such a thing?"

"For the cooking maybe," Mitch said, spitting out a chunk of gristle.

An uncomfortable silence followed. Suddenly it became apparent that one of the quieter members of the group had been married ten years and another was engaged to be married at the end of fire season.

"Ah, it's probably not so bad," said Wally Humphries, the engaged one.

Tony Pastro looked up from stirring his coffee.

"What would you guys know about it anyway? Asking you guys about marriage would be like asking Peter Pan about life insurance."

The rest of the jumpers held up, their minds wrestling with the concept.

"My wife calls it the Peter Pan theory," Tony went on. "Most smokejumpers live like Peter Pan, not wanting to grow up, thinking they can live forever flying through the sky, adventure after adventure. As the years go by, Tinker Bell loses some of her tinkle, and the croco-dile comes closer and closer, the ticking of the clock getting louder and louder. Meanwhile, Peter Pan just keeps on flying around in never-never land."

A silent moment followed. Leaves rustled in a light breeze.

"Well," Bell finally conceded, "I don't figure to worry about it much. Lots of things happened to me that I didn't choose. I've been with some good women, but I also knew that I wouldn't make a decent husband, and so I got myself away from them.

"And as for crocodiles, ticking or otherwise, they'd better not come around and start messing with me . . . scaly, reptilian, high-eyed bastards."

"What was that song in the movie?" Kubi chuckled. "Never smile at a crocodile?"

"There won't be no smiling to it," Bell protested. "I'll be killing them with my bare hands, breaking their jaws, stomping on their heads, and popping their eyes out . . ."

"Bell," Mitch interjected. "It's just a theory, man. There's no need to get uptight."

"Who's uptight," Bell shouted. "I just don't like theories, that's all. Especially about things like reptiles and women."

"Pass the flat nose," Wally said.

"It's not a bad theory," Mitch reflected, "but fire season's no time to go gettin moose-eyed with Tinker Bell. This job'll eat your love life just like the crocodile did the guy's watch—hand and all."

I went over to take a look at Al's eye.

"The area around the sliver is bloody," I said quietly. "It's worked its way in deeper, too."

I was about to ask if he wouldn't reconsider going in to have it checked out, when he told me that he was ready to go. I called Troop and told him we were sending Al to the helispot for an early demobe. Troop told me to bring everyone up. He wanted to hold a meeting.

At the meeting we discussed strategy for the upcoming day. Troop planned on setting up a portable tank on top of the ridge and using a helicopter equipped with a heli-bucket to keep it full. The two crews would mop-up with hoses and Pulaskis. Eagle decided to keep Troop plus three other jumpers on the fire; the rest would leave that afternoon.

Jumper rules are choice from the top, bone from the bottom, meaning we go by order of the jump list. Those at the top can choose to stay or go. If enough don't volunteer to stay on the fire, then those who jumped last have to. If you wanted to go but are forced to stay, then you got boned from the bottom or just plain boned. Being one of the first to jump, I got to choose. Thinking of Sally, I decided to leave.

Maybe things would slow down. Maybe I could spend some more time with her. Aside from that, my knee hurt constantly, and a rest might do it some good.

"What about the bear?" Kubi asked.

"Give me a Pulaski," Bell said.

"He's hungry," Kubichek said. "He wants *all* our food."

"Well, he's not going to get *all* our food. He's not getting any of it. *This*," Bell said, brandishing a Pulaski, "is what he's going to get."

"He's full grown," Kubi said. "He must weigh five hundred pounds."

"Too bad," Bell snarled. "I've had it with these shit-haired bears that go around acting tough. It's time they found out who's boss."

"He ran those other jumpers off this fire," Troop said.

"He chased me as soon as I started yelling at him," Kubi wailed.

"That may be, but his chasing days are over. When it comes right down to it they're nothing but stinking little pig-eyed hairballs." Bell pursed his lips, his forehead pinched into a frown.

"Bears," he said. "There's only one thing they understand. Physical force!"

"How many guns do we have?" Troop asked.

Among the group there were three pistols, a .44 Magnum and two .357s. Troop assigned the jumpers with guns to work the end of the fire nearest the jump spot. A 12-gauge shotgun had been ordered from Eagle. The BLM's unofficial policy concerning bears is simple: Don't shoot them unless they're eating you. Instead of killing them, we generally give them all the room they want, moving our camps and keeping out of their way. If that doesn't work, chasing them off with helicopters usually does.

The grizzly bear (*Ursus arctos horribilis,* or "the horrible one" as we call it) is generally not a problem. As potentially dangerous as they are, we have few encounters with them. We may see one as we circle the fire preparing to jump, but by the time we're on the ground and finish dropping our cargo, it will have typically moved out of the area.

Black bears, on the other hand, tend to hang out, dozing on sunny hillsides, eyeing camp and lifting their noses now and then in our direction. Some have simply walked into camp, dug into food caches, and made themselves at home. Once they can no longer be intimidated by yelling, rock throwing, or the chain saws, trouble is inevitable.

Lynn Flock, an Alaska jumper, once had an interesting encounter. He and his crew had jumped a fire, worked it late into the night, and then returned to the jump spot. They were tired, had eaten quickly, and left food boxes scattered around. After putting up their hootches, they immediately crashed. Rich Halligan and Lynn had shared a cargo chute hootch. I don't know how long they'd been sleeping, but suddenly Lynn woke up thinking he'd heard a noise. He was so groggy, though, he couldn't tell if he was awake or dreaming. There came this steady push on the chute near his feet, and he thought it must be one of the jumpers. Suddenly, a set of huge claws punctured through the fabric. Next thing Lynn knew, a black paw had ripped a hole and filled it with a hairy black head.

"That got us fairly excited," Lynn told me. "I bolted up in my sleeping bag, Rich bolted up in his. The bear looked at us. We looked at him. He grabbed the end of my sleeping bag and started pulling it out the hole. I wiggled out about the time the bear had it halfway out."

After a very long moment, Lynn and Rich heard the bear messing around with another hootch. Lynn poked his head out. Ten feet away the bear was standing on its hind legs beside Brian Fitzsimmon's hootch. Fitz had stretched a cargo strap between two black spruce and tied his tent and rain fly to it. This crazy bear held out its paws and belly flopped on Fitz's hootch. If the cargo strap had broken, the bear's great bulk would have crushed Fitz. But the strap held and suspended the bear high-centered, astraddle the whole mess.

Fitz woke up, annoyed. "All right you fuckers . . . knock it off. This is no time to go foolin' around."

Lynn yelled back, "Fitz, it's a bear. Bear in camp!"

Fitz started screaming. "Get him off. Get him off meeeee!"

The scream stopped abruptly, and there followed a complete silence. Then the bear came padding back toward Lynn and Rich's hootch. Halligan remembers thinking, "Whatever else may have been happening in the world at that moment, I knew only the following: A bear had torn up our hootch, gone next door and killed my neighbor, and was headed back our way with, I assumed, homicidal intent."

As Lynn and Rich grabbed their clothes and tried to dress, Lynn's head was bumping the side of their hootch. The bear came over and gave it a swat. They both screamed. Halligan tore out of the tent and immediately ran into a tree. The other jumpers had gotten up and started running around frantically. One went for the chain saw, some grabbed fusees.

The bear headed back for the mess he'd made of Fitz's hootch, then turned to face a gang of strange hairless creatures waving fusees. Hot slag from the fusees started dripping on their feet, which made them hop around, stepping on more hot slag, cussing at the top of their lungs. Small fires started from the slag. Next thing they knew, their camp was on fire and they were running naked in circles.

About then, Emmett Grajalva showed up with his .357 Magnum—Emmett, of all people. Emmett, the man with a full-on phobia for bears, a man who talked obsessively about them and feared them more than anyone else on the crew. For Emmett it was bear Armageddon, a last hateful showdown between man and bear. Bear *High Noon*.

Emmett took aim at the beast but was shaking so hard he was afraid he might shoot Fitz instead. Finally, in terror, he turned and fled, ran smack into a tree and broke his nose. The other jumpers were fighting fire, knocking everything into disarray, and continuing to burn their feet. As soon as they had beaten down the flames, they paused to take a good look. Their camp had been reduced to a pile of incinerated wreckage, while the bear had gone his way, ambling easily out across the hillside as if the whole thing had been no big deal.

At ten o'clock a helicopter arrived with the first load of supplies from Eagle. The pilot told us that the first eight jumpers would be leaving on the next flight—sometime around noon—and the rest on the load after that. I talked the pilot into taking Al into Eagle right then so that he might be able to connect with an early ride back to Fairbanks.

Eight of us were gathered up and waiting with our gear bags on the helispot. Bell had made a cursory search of the area for the bear, returning to the helispot, Pulaski in hand.

"Worthless, pig-eyed, thieving cowards," he said. "Hiding out in the brush, sneaking around, afraid to show their faces."

"There he is," Kubi shouted.

We all jumped to our feet. The bear emerged from the woods north of the helispot, then hesitated, eyeing a pile of fire packs. Bell began walking in the bear's direction. The rest of us moved in behind. As Bell crossed the center of the helispot, he screamed, jumped up and down, and made fake charges. The bear lifted up on its hind legs, its head now eight feet above the ground. The rest of us yelled and threw sticks. The bear dropped down on all fours. Hair lifted up on its back as it lowered its head and began swinging it back and forth, all the time eyeing us steadily. Bell charged in a primal rage. Black and shiny, the bear wheeled and broke into a lope back into the trees. Bell threw his Pulaski at it and then commenced laughing like a maniac.

Stooped over, he slapped his knees repeatedly while staggering in circles.

"Come back here," he wailed. "Come back here and fight, you big fat ass."

I've seen Bell pretty happy at times, but I've never seen him that happy. Soon we were all celebrating with him, jumping around, high-fiving, and filling the air with *ahh-uhhhs* and laughter.

A Twin Otter sat on the runway in Eagle. So did Al. As soon as we landed, he got up and headed our way, hunched over, struggling against the rotor wash of the helicopter, a dirty bandanna holding a fresh white bandage over his eye. Al slid open the side door and after transferring our gear over to the Otter, we were off the ground again. While the plane was still climbing, I looked around and saw jumpers crashed in every available space. Al lay sprawled atop a stack of fire packs, ghastly and wounded looking, a dark-colored fluid oozing out from under the bandage. Cargo had been stacked to the ceiling—all except for one fourteen-inch space about six feet long. I crawled up, squeezed in, and quickly fell asleep.

Inside the shack I checked the jump list. I made the fourth load—down far enough to have an excellent shot at the night off. Four of us from the Canada load were told to go ahead and eat but to hurry back. Al disappeared into the emergency medical tech's office.

At the cookhouse we pigged out on pork chops, homemade bread, apple pie, pineapple ice cream, and watermelon. Filthy and happy, we ate and talked with our mouths full. A cook came to our table and said I had a message—Sally had just called and wanted me to call her. I went to the phone and dialed her number. Yes! I could pick her up about 10:20 at the Captain Bartlett. I ate another piece of watermelon, wrapped two chops in a napkin, and stuffed them into my shirt pocket.

Back at the shack we discovered that three loads had just left. Tom Romanello was the box boy running the operations desk. "Hustle up and get your shit on the racks," he hollered. "You guys aren't just the first load in Fairbanks. You're first load for the entire state."

I was stunned. I felt a serious urge to throw my helmet against the wall. "I can't fucking believe it," I said quietly, cupping my hands to my face. At least we ate, which was more than Bell, Mitch, and the others that were already on their way to another fire could say.

With three loads rolling, the jump list showed me first man, first load. Everyone hustled off to get fire ready. I stepped outside the doorway of the ready room and screamed, *"Bullshit!"* into a dismal hazy emptiness, then stood there glaring out across the airfield.

The runway lights had been turned on so pilots could see to take off and land. The air smelled sweet and tangy and dangerous. I was tired. My knee throbbed in pain.

"Please," I mumbled. "No more fire today."

Frustrated and somewhat ashamed, I stood there trying to pull myself together. I wanted to see Sally, but instead I was going to have to go back into the shack, get my gear ready, and most likely jump another fire, take charge of a load of young jumpers, and brush-ape it up all night, and then in the end call the whole thing fun.

I looked over the jump list. The others with me on the first load were young and relatively inexperienced. There was Kubichek, a snookie; Togie Wiehl, a rookie; Morgan Whipple, a four-year solid hand from Missoula; and four other rookies and snookies from Missoula, all new to Alaska fire. I'd seen Togie very little since the fire at Rampart, but from what I'd heard, he was having an all-star season with the most fires jumped and the most overtime of anyone on the crew. Fire-line stories were already circulating about the heroics of the amazing "Rogue Togue."

I dumped my gear on the floor and began sorting through it, making ready. That done, I checked the jump list again. Seiler's name tag had been put down into the lower right-hand corner of the magnetic board where it joined four others: Robbins, Brown, Cochran, and Boatner—all listed as "cripples." I went to the rest room and had barely settled into my personal preflight duties when the siren sounded.

"Hey, Romo," I yelled to the box as I started putting on my jump gear.

"Call the Captain Bartlett. Leave a message for Sally. Tell her I can't make it."

"Got it," he said, quickly bending to the notepad on his desk.

I waddled out onto the ramp, encumbered by sixty pounds of protective gear, two parachutes, and a twenty-pound PG bag. A PB4Y air tanker began takeoff, the power of its four engines pounding and echoing around the buildings. As it rolled past the shack, the hair tingled on the back of my neck. I once heard Jack Dyer, an old-time 4Y pilot, say, "When it starts to hurt my ears, that's when I know I've got enough power to get her off the ground."

Our own plane, the familiar *Jump 17,* finished running up its right engine and began cranking left. Like a mother duck, I led my fledgling jumpers out onto the flight line and into the noise.

We came to the eastern edge of Minto Flats, with its waterways twisting and bending and turning back on themselves in gray oxbows and sloughs, a passing reflection of violet. The Tolovana River.

Our engines settled down to a soothing drone. Feeling drowsy, I dropped onto the floor, stuffed in my earplugs, rested my head on a couple of cargo chutes, and looked at my crew. They seemed so young. For two, this would be their first fire jump in Alaska. For another two, it would be their first fire jump anywhere. I knew I could count on Togie and Kubi, but they'd just come off the border fire and were tired. Whipple was a definite asset; a young and tough Zulie—a Missoula smokejumper—we'd been on fires together before. But of the Missoula rooks and snooks, I had no idea. They looked strong enough all right, and they were fresh and eager for action. They'd just arrived with the most recent booster crew from the Lower 48. One had a sticker on the side of his helmet that read GOTT MITT UNS—God help us.

Shortly I felt a tapping on the toe of my boot. Through my face mask I could see Togie waving a note.

Fire B-595 Taylor—Fire Boss
5 Acres—black spruce—totally active——CRITICAL
2 miles south of Coldfoot Camp (Alaska Pipeline)

Tucking the note in my shirt pocket, I rolled over and tried to go back to sleep. Again I was awakened by the dutiful rookie, who must have thought my attitude a bit nonchalant. Another note.

B-595 20 acres—rolling, hot, thick spruce.
Air Tankers: T-126 Fort Yukon
              T-84   Fairbanks
              T-67   Fairbanks (all en route)
Air Attack: 921 ST (Sierra Tango)—Fort Yukon.
              Estimated ETA—2135 hrs.
FM—Gold: Channel 6
Structures threatened—(mining claim?)

Pulling off my helmet, I reached up on the side of the fuselage and took down the spotter's headset.

Leonard Wehking, our spotter, was talking to the pilot. "Well, let's see here. Hmmm . . . that over there . . . that must be Hess Creek . . . Don't you think?"

"Hmm," Mark, the pilot, said. "We must have missed Livengood then."

"Yeah," Leonard agreed. "I guess we did . . ." He chuckled softly. It ain't much. In fact, I can't remember just what it looks like from the air."

Leonard, a balding red-haired Utahan, is known among the jumpers as Left Bank. He jumped a fire once by a big river and wound up landing on the opposite side from everybody else. Fearing he'd drown if he landed in the water, he'd turned and run with the wind to dry ground.

This created a logistics dilemma: seven on one side, one on the other, and the river too wide to cross. The fire boss had to decide what to do about Leonard. Our fire packs supply two people. If he dropped Leonard a fire pack, that would short them supplies for one man.

"Go ahead," the fire boss had told the spotter, "drop one to the left bank."

That's how Leonard got his nickname. For two days, the seven fought fire while Leonard constructed an elaborate camp, ate lots of food, took long naps on his two sleeping bags, fished, hiked, and waved encouragement to his friends across the water.

"Left Bank," I asked over the headset, "are you lost?"

"Well, well," Leonard muttered, apparently pleased. "If it's not Murry Taylor, my old jump partner. The way you was a layin' around back there, I thought you was dead."

"What's the deal on this fire?"

"Didn't you get the notes?"

"I just wondered if you knew anything else."

"That's it! What's in the notes is all I been told. They don't know the source of the report, but they're pushing the panic button 'cause it's so close to Coldfoot."

"Leonard?"

"Yeah?"

"Leonard, I'm your old jump partner, right?"

"Murry Taylor," he crowed. "My *old*, old jump partner."

"And my good buddy, right?"

"I was just tellin' Mark here how you and me was such good buddies."

"You wouldn't put me on some god-awful fire, now would you, Left Bank?"

"Not on your life! Hey, I had a fire up here in '89. In two hours we had the thing out and were eating homemade pie at the Coldfoot Cafe."

"So when we get there, can I tell them Left Bank sent us and just go ahead and put the pie on your tab?"

"Damn! What a pig," Leonard cried. "Here you're making all this overtime, and you're still too cheap to buy your own pie. No wonder you're havin' trouble finding yourself a woman."

I was searching for something nasty to say about Leonard's love life when he broke in.

"Hey, big spender, we're about fifteen out. Better go ahead and do your checks."

After completing our buddy checks, I heard Mark, the pilot, on the headset.

"Oh, oh," he said softly. "That doesn't look too good."

"Oh, no shit," Leonard said.

"OK, you two," I interrupted. "You guys are marginal as pilot and spotter, but your comedy's even worse.

"No! No!" Mark insisted. "We wouldn't do that."

There followed a pause, and I thought I heard some background giggling.

"Bullshit," I said. "You're jacking me around. Probably Leonard's idea."

No response. The silence grew. Then I heard Leonard.

"Yeah," he said to no one. "That's one hell of a smoke column." Then to me. "Hey, pardner. Sorry about this, but I hope you didn't have your heart set on eatin' any pie tonight."

Minutes later we were in among the mountains, dropping down for a closer look. Two fires burned side by side, out of control on a broad incline at the base of a tall mountain. One was about twenty acres, the other about five. A cluster of buildings stood directly in their path a half mile out in front. Five miles to the east, I could see the Alaska Pipeline, the Haul Road, and the buildings at Coldfoot Camp. The rookies had that vacant, scared-shitless rookie stare as they watched the heads of both fires flare up into fifty-foot flames. The fires ran parallel to each other, with about 250 yards of unburned tundra and spruce between them. The perimeters of both fires were fully active, producing flames from six feet at their tails to twenty feet along the flanks.

The lineup for our final pass would be between the two almond-shaped fires. The jump spot was at the mining camp and lay directly in front of the largest fire, partially hidden beneath heavy smoke. Just beyond the buildings, there was a junction of two creeks and a small airstrip grown over with willows. To the rookie eye, it must have appeared that we would be jumping in front of two fires that were rapidly closing in on each other—which, in fact, was sort of the way it was.

Mark and Leonard worked out the pattern for the drop. Then Leonard came crawling back over the stone-faced rookies. He pulled open the door and the air roared in. After the streamers, Leonard turned with a rather grim look on his face and called me and Togie to the door.

"I don't think this is as bad as it looks," he yelled. "You've got to get to those buildings. We don't know if there are people there or not, but that's your first priority. Did you see the airstrip?"

We did.

"The spot's at the end of it by the buildings. Just don't land up on the bluff between the fires. If the winds get tricky, go long, and stay wide of the big fire. You'll have to drop down through that smoke there beyond the bluff."

We nodded our heads, and Togie and I got ready. The plane pulled around on final. Dropping down into the door, I completed my four-point check and glanced down between my legs at the two fires. Leonard gave me the slap. I pushed off, tumbled out into space, and in an instant was accelerating to ninety miles an hour. The wind wasn't that strong up high, but it swirled wildly around the head of the main fire. Descending into the smoke, I lost the spot. It was scary as hell, caught in the clutches of gravity, unable to see, but coming down just the same. Once under the smoke, I could see the mining camp and the junction of the creeks. On my final approach, a hundred yards out, I looked up and saw Togie right above me. Square parachutes act like kites when they collide, spiraling into one another, snarling, and likely taking each other down. I knew he couldn't see me since I was a little behind him, so I turned left and gave him the right-of-way.

I'm not certain what happened next, but it seemed like the wind picked up and I couldn't gain back the ground I'd lost. I was heading straight for the creek, which appeared to be about thirty feet wide, deep, and treacherously swift. At the last second, I reefed down hard on my right steering toggle, swung to the side, made a low, radical turn, and crash-landed on the wrong side. I could almost hear Left Bank laughing his brains out up in the plane. What a sorry sight I must have been wading that creek, water to my waist, holding my radio over my head—the old salt missing the spot and having to face a bunch of rookies and snookies who had all landed where they were supposed to.

"Hey, there, old-timer," Leonard sang out over the radio. "It's a little early to be takin' a swim, ain't it?"

"I've had it with this fuckin' job," I yelled into my radio. "The fuckin' rookie ran me out in the brush."

"A likely story." Leonard laughed. "Don't worry, I won't say a word about it back at the shack."

"Have a nice day, Left Bank."

"Will do, Right Bank," Leonard said, laughing more.

First thing I did was to assign Togie to check the buildings. There were no people; the camp was abandoned. We retrieved our cargo, set up a pump, and stretched out five hundred feet of inch and a half nylon hose just as the main fire crested the bluff above us. A wall of flames came licking over the rim, raining sparks and firebrands down on the camp.

"Togie," I yelled. "I'm leaving you here with three Zulies. Keep the pump going. I'm taking the rest up the bluff and along the right flank. Once these buildings are secured, you stay here but send the others our way."

I had complete confidence that Togie could handle the problems of protecting the camp—even though he had run me out of the spot.

After lining out Kubi and the two others, we worked our way up the steep bluff, knocking down spot fires and scratching hand line as we went.

Six days later I would submit the following narrative to Bettles dispatch as part of my official report for the Coldfoot fire:

July 3.   Jumped fire B-595 (two fires) three air miles west of Coldfoot Camp. The East Fire burned 40 acres; the West Fire, 8. The first load of jumpers jumped at 9:50 in the area of the structures and set up pumps and hose. A second load jumped the West Fire at 11:45 P.M. Buck Nelson took charge of that load. Nelson and I conferred over the radio about our plan for the night. After securing the perimeter, we left the fire line at 3:00 A.M., returned to the jump spot, got a quick bite to eat, and went to bed.

July 4.   By 9:00 A.M. I had the first load back on the line mopping up perimeter smokes and hot spots. At 2:00 P.M. an undetected spot fire across the creek took off on Buck and his crew. The East Fire was kicking up as well, keeping us hustling—the wind constant, warm, and dry. Buck suggested that I radio for air support. We spent the next two hours running from one flare-up to the next, knowing that if either fire was lost, it would pose a threat to those caught in between.

Two air tankers arrived at 4:10 to begin a typical afternoon air war. They made their last drop about 6:30. By midnight everyone was back in camp, dead tired, crowding around cooking fires. Both fires had been stopped.

July 5.   Bettles wanted as many jumpers as I could spare. We negotiated and struck a deal. I would keep seven jumpers, including Nelson, and release the other eight. Then I placed the following order:

1. 2 16-man EFF crews with tools
2. 40 cases MREs
3. 40 5-gallon cubitainers, drinking water
4. 8 extra Pulaskis
5. 18 Fedcos backpack pumps
6. 2 Mark III pump kits—complete
7. 4,000 ft. $1\frac{1}{2}$-inch nylon hose
8. 4 gated Y's
9. 4 extra $1\frac{1}{2}$-inch nozzles
10. 1 case plastic garbage bags
11. 1 bundle burlap bags
12. 6 5# cans coffee
13. 10 cases juice—12 oz. cans
14. 24 rolls toilet paper
15. 1 roll EFF cord
16. 3 rolls fiberglass tape
17. 2 large camp tarps
18. 12 King radio batteries

July 6.   Continued pumping water on fire perimeter, patrolled for spots. Notified Bettles of plan to demobe the Allakaket #1 crew, July 5 at 6:00 P.M.

July 7.   Completed mop-up. Broke down pump show. Continued gridding both fires. Demobed Allakaket #1. Notified Bettles to go ahead with plan for total demobe on the 8th.

July 8.   Final gridding—both fires. At 2:00 P.M., started the Allakaket #2 crew shuttle direct to Allakaket. At 6:00 P.M. packed up jumper camp and demobed B-595 completely with final helicopter shuttle into Bettles.

General Comments:

1. Excellent support from Bettles dispatch. Problems getting fresh food. If we'd gotten weathered in, I would have had 40 people with nothing to eat.

2. The mining camp turned out to be a bunkhouse, a kitchen, and a couple fallen down storage sheds. The area is trashed with all sorts of discarded mining equipment: pipe, scrap metal, cables, chains, lumber, plus a 1940s vintage army four-wheel-drive Mule. The kitchen has been ransacked by a bear that came through a window, cut itself and bled all over a large table. From the looks of the table, he spent a lot of time trying to lick up his own blood. The kitchen and everything in it is a total loss.

3. Records found in the bunkhouse corroborate the dates found on a few magazines. Apparently the last mining took place the summer of 1985, when an activity log shows them taking out $85,000 in gold dust and small nuggets. The note I got flying to the fire indicated that this fire was burning in Critical. Is that because of Coldfoot Camp? This deserted mining operation wouldn't seem to warrant it.

Murry A. Taylor, Fire Boss

There were a few matters I didn't mention in the fire report. Again I'd had the good fortune to end up at the southern edge of Gates of the Arctic National Park. Large storms stalked the late afternoons, their resounding thunder pounding all around as they moved up canyons and across ridges, where they finally rolled away to the north. Once the storms had passed, the evenings grew calm, and the clouds parted to low-angled sun, rainbows, pastel purples and pinks, while our camp smoke floated electric blue in the air above the creeks.

The bunkhouse was neat and tidy. Its windows were covered with screens and its beds held soft, clean mattresses. The sleeping was excellent, with the sound of rushing creeks outside our door. During the evenings we enjoyed a quiet camp life, cooking, eating, telling stories, and sharing a nip of Mr. Daniel's Tennessee whiskey. I had ordered fresh food the same day we jumped. We thought the food would come on July 4, and were disappointed when it didn't. Surely, we thought, it would come the next day.

When the Allakaket crews arrived, I found out that they'd come from another fire and had no food, so we gave them what had been left by the other jumpers. By midday of July 5, our fat boy boxes were badly ratted. Only crumbs remained of the Ry-Krisp—no more peanut butter, no jam. The canned ham was gone, the smoked oysters, too. Only a few instant oatmeals (both with and without weevils) were left, along with two cans of green beans. Radio contact with Bettles was shaky. From all I could tell, our fire was a priority for fresh food, but each day Alaska was getting twenty new fires, and there was a shortage of aircraft. Later that night it started.

"We're outta food," Togie said, rummaging through a fat boy box that had earlier been used for garbage. Togie grinned at me, his eyes kind, not accusing, but hungry nonetheless.

"This isn't funny," Kubichek moaned. "We're gonna starve."

"You ate all the fuckin' candy bars, you hog," blasted a Missoula jumper, an acquaintance of Kubi's.

"I did not!"

"You started rat-holing the minute we got here."

"Bullshit!"

"Everybody knows you're a flaming rat-holer. Don't deny it. Every morning you come to camp with chocolate on your face and that goofy sugar-blues look."

"Kubi always looks goofy," Buck was quick to point out.

"I'm being framed," Kubi squalled.

"The hell," the Missoula jumper said. "Every fire I'm on with you, there's a trail of candy wrappers all the way to your hootch. You've probably gained twenty pounds since this bust started."

"I can't believe it," Kubichek said. "My own friends, my own bros!"

"Bros my ass," said the Missoula jumper. "You're like a sheep-killing dog. If we don't get some food in here pretty quick, *your bros* are gonna roast your fat ass right over this fire."

As the man in charge, I felt particularly uptight about the food situation. Terrible things can happen during food shortages. Sometimes big guys jump little guys and take their provisions. Once, on a fire in the mountains southeast of Tanana, Don Bell went berserk at the sight of a canned pound cake when I pulled it out of a C-ration box. He was on me like a cat. Down the hill we rolled, knocking over two tents, Bell practically breaking my arm as he wrenched the cake away. Back at camp, he opened it and ate the whole thing in one bite, cheeks puffed out, choking bug-eyed.

John Rakowski, "the Legendary Fucking Rake," or simply "the Legend" as he liked to refer to himself, got back at Bell. The same night, before going to bed, the Legend found a new-looking ration box that listed pound cake as the dessert item. He opened it, took a shit in it, and put it where Bell would be sure to find it first thing the next morning.

Bell was big and confident and knew no one could take anything away from him, so when he found the new C-rat box, he proudly brought it over to the campfire to show it off. "Ha, another pound cake," he announced. Still half asleep, he tore off the top of the box and rammed his hand inside.

"*Shit!*" he screamed, jumping back and throwing the box in the air.

"Damn it! That's not cake . . . that's *shit!*"

Don starting cussing, rubbing his hand on clumps of grass, snorting furiously.

"You know," Rakowski said, with a giveaway grin, "these rations are getting shittier all the time."

The next thing we knew the Legend was bolting off down the mountain with Bell hot after him.

The idea of roasting Kubi reminded me of other times when the subject of cannibalism had come up. Up on the Kobuk River in 1977, an entire crew of sixteen had been lost by Galena dispatch for nine days. They had caught a few grayling and picked blueberries, but appetites were keen. The topic of who might have to eat who came up, which led to a survey of prospective candidates. A few had been immediately rejected—Bob Quillin, for one. Although widely renowned as the Stewing Chicken, Bob had not been a popular choice.

"Too much bone. Stringy meat," his colleagues had reasoned.

Then, there was Erik the Blak, considered inedible under all but the most dire circumstances.

"Too many toxins," the crew had complained, recalling the Blak's love of cheap whiskey and Copenhagen chewing tobacco.

On the Coldfoot fire, it didn't take long until we had a consensus. If it came down to anyone getting roasted, it would have to be Kubi, a plump and proven rat-holer. Besides, the rookies present were skinny, and one was a woman and thus was protected by male instinct from being considered as food. When these points were made to Kubi, he immediately rushed off to his tent, got some fishing line, came back, put a hook on, ran out to the creek and began thrashing the water with a willow pole.

By the evening of the sixth we'd been out of food a full day. Worse, we had lost contact with Bettles altogether. Togie and the two Missoula jumpers made lures from tin cans, cut willow poles, and set out fishing, but all they caught were their lines in the brush.

Clouds formed near the mouth of the canyon, making it doubtful that a plane could get to us anyway. At eleven, Buck and four others decided to turn in. Togie, Kubichek, and I had lost hope, too, but we decided to sit up awhile longer. No sooner had the others gotten quiet in the bunkhouse than the silence was broken by a radio transmission and the sudden roar of an aircraft.

"Fire B-595, this is *Jump 17*."

Those of us at the campfire jumped to our feet. A commotion in the bunkhouse. The boys scrambled out naked and began milling in circles looking up. I grabbed my radio.

"*Jump 17*, this is 595. Welcome."

"Good evening, gentlemen. Got a little surprise for you. Sorry about the delay. Too many fires in front of you." It was Dalan Romero. "Beautiful country you've got up here."

Jump 17 circled camp as I explained to Dalan about the food to be dropped to the Allakaket crews camped up the side creek. Moments later, *Jump 17* dropped down out of a blazing sky and roared over at two hundred feet. Pass after pass, bundles tumbled from the plane to be quickly snatched up under brilliant orange and white parachutes. Four bare white butts raced after them, arms waving in pursuit of our newly arrived manna from heaven.

Jump 17 finished dropping to the other camp and then began her climb back to altitude. When Dalan came back on the radio, I could hear the aircraft engines humming in the background.

"OK, Murry, that's it. We've got to go. We're burning flight time."

I thanked both Dalan and the pilot for hitting the sandbar instead of the creek and tried to express how much we appreciated the delivery.

By midnight our campfire was scattered into a perfect bed of coals and a fresh pot of coffee was brewing. Overhead, a yellow-green sky ushered in the new day as the aroma of barbecuing steaks, hot dogs, and fried potatoes and onions filled the air.

July 8

While waiting at Bettles for an airplane to take us to Fairbanks, I went over to dispatch to turn in my fire report. The two female dispatchers were friendly, helpful, and big-breasted, so I stuck around for an extra cup of coffee and some visiting. We chatted amicably, passing a few minutes talking about my fire, the current fire season, and fire busts in general. When I got back to the airstrip, I found the bros gathered up on the loading dock of the smokejumper cache listening to the two jumpers who had originally jumped Fire B-405, the border fire just east of Eagle.

"What a bone job," one moaned. "We didn't have any luck on that thing at all. The fucker was only about an acre when we jumped it, but it was dry as hell. We worked through the first night and all the next day until midnight. By then it was lookin' pretty good, so we decided to head back to the jump spot and get some sleep."

"Right after we got there," groaned the other, "that piss-ass bear showed up."

"Showed up, shit," rebuffed the first. "*Came back* is what he did. We ran him off, but he'd already eaten, bitten into, or slobbered all over everything."

"That son of a bitch," his partner muttered. "We didn't give a shit. We were too tired. So we just went ahead and ate what we could, put up our hootches, and went to sleep. Next thing you know, there's these god-awful snorts and grunts. I look out my tent and the son of a bitch is back into the food again. I figured, I'll fix his ass. So I get up and yell at him and he just lays down, looks at me, and keeps on eating and snorting and licking his lips. I got a big stick and went over and waved it at him. He kept watching me out of the corner of his eye. I started beating the ground with the stick. That's when he jumped up and charged. Shit! I ran like hell. He didn't chase me far though, just away from the food.

"By then my jump partner was up, and he says, 'What are we gonna do? We can't just let him eat everything.' We had a time with that nasty bastard, I tell you. We kinda lived with him for the next day or so, then the plane from Eagle flew over, and I told 'em we needed water for the fire, more food, and a submachine gun to blow that motherfucker to pieces."

He stopped a moment to remember.

"Bastard," he said. "We threw the worst torn-up stuff over in a pile close to where he was and tried to keep the best stuff for ourselves. But he didn't go for it. He wanted both piles. We couldn't sleep without one of us keeping watch. Then, check this out. While we're out patrolling the fire, he knocks one of the tents flat, tears into it, slobbers all over the sleeping bag, and takes a big shit on it."

"Then," his partner added, "the plane came back and told us we couldn't get any food or water. Too many fires, not enough planes. They suggested we leave the fire. Check it, then leave."

"That's what we did. But we took every last bit of garbage with us, so we screwed that bear in the end."

"Well, it wasn't a very big screwing," the partner objected. "There wasn't that much left. Truth is, he already had us plumb screwed."

Hearing a plane, we looked up to see one of the new blue and white BLM Sherpas approaching from the south. Gathering up our gear, we carried it to the edge of the strip. The Sherpa pulled up and parked right in front of us. We loaded our gear on board, crawled in, took off, and settled down for the flight back to Fairbanks.

Ten minutes later there was a stirring at one of the windows, fingers pointing and eyes bugging until I couldn't stand it anymore and had to look for myself.

"What is it?"

"Down there," Kubichek pointed. "In the trees by the bend in the river."

There in a thicket of spruce was an aircraft wing. Then I saw the tail, and the whole picture came back.

"That's that Helio-Courier that went down last summer," I said. "We almost jumped it."

"There's the rest of it in the meadow," Togie shouted. "All burned to shit."

"They think it exploded in flight," I said.

Our eight-man load had rolled out of Fairbanks initially in search of a missing helicopter but then was diverted to a crash. Apparently the pilot had radioed Bettles that he was twenty miles southwest with his engine on fire. When we arrived on scene the fuselage was still smoking in the middle of a meadow about a quarter of a mile from the wings and tail. We made a low pass, and I thought I saw what was left of a human torso in the cockpit. We'd been about to jump when the state troopers radioed us to hold off—they would take it from there.

The fire was burning in a let-burn area, so we turned back toward Fairbanks and left the troopers to the grim task of recovering the remains of the three dead.

Alaska has more than its share of aviation mishaps. Only California and Texas, with much larger populations, have more airplane crashes. It stands to reason—the great distances, the radical and often unforecasted weather, the manner in which aircraft are used and misused. In the north country, aircraft are often flown as casually as people drive automobiles elsewhere. I once saw a man in Fort Yukon rush out of his cabin, jump in his Cessna 185, hit the starter, spin the tail around, and take off down the runway in a cloud of dust. No run-up, no oil-pressure check, no magneto checks, just crank up and fly.

In the mid-seventies two of BLM's DC-3s blew engines. One lost its left engine on takeoff out of Tanacross, the plane fully loaded with jumpers and gear. That had required an immediate turnaround and landing. The other plane, which I was on, was taking off out of Beaver, bound for Fairbanks. We'd been on standby in Beaver for three days, and the pilots wanted no more of it.

"Keep your parachutes handy, boys," the captain yelled. "This old girl's going to Fairbanks."

DC-3s fly fine on one engine, so it wasn't that big a deal, really. We always flew with our harnesses and our emergency chutes close by, so the worse that could have happened would have been a free jump. The pilots didn't have any chutes, but that didn't matter much since most of them would rather crash a plane than jump out of it.

Then there were the Volpars, for twelve years the primary smoke-jumper aircraft in Alaska. In 1976 I was on a load preparing to land at McGrath. On an extended downwind leg I saw that the pilot had become preoccupied with something. The landing gear was jammed and wouldn't drop. The pilot needed to get rid of us fast and head for Anchorage, where he could belly it in on a runway with emergency foam facilities. The wind was blowing fifteen to twenty miles per hour and there was no time to throw streamers.

As the plane angled in over the McGrath runway and banked into a tight turn, we bailed out—all eight. We landed all over the place: Bill Recinos in a waist-deep backwater slough, two others in a dense stand of cottonwoods. At the edge of a clearing, I came down backward, slammed into the ground, and was knocked unconscious. I came to sprawled over a log with my legs numb and my chute tangled in the willows.

By then, the plane was on its way to Anchorage. Heading over Rainy Pass, the pilot doubted he had enough fuel to make it through the Alaska Range. There was one parachute on board, and the pilot tried to talk Rob Collins, the spotter, into jumping. Rob declined. Shortly after that the pilot became furious, handed the controls over to Rob, climbed up on his seat, and began kicking the center console. Suddenly the landing gear mechanism came to life, groaned, clunked, and slammed the gear down into position. They flew back and landed at McGrath, where the ship was grounded.

On another Volpar, during the great 1977 Bust, the reverse thrust mechanism on the right propeller malfunctioned upon landing in Galena, sending the plane with its pilot, spotter, and four jumpers on an unexpected spree into the willows alongside the airstrip—a little side trip that ripped off the landing gear, mangled the props, and left the ship lying flat on its belly and filling with smoke as the jumpers came flying out like popcorn.

In August that same year, we almost lost another Volpar on a patrol flight in the Alaska Range. Davis Perkins, a former Alaska jumper, wrote the following in his journal:

> Of most significance is an experience that happened on August 8, in Volpar 46 Victor, flying on the deck in a valley leading into the Alaska Range. It was a beautiful flight until we made a right turn up a valley. Within seconds everyone on board knew we were in trouble. The valley suddenly ended in a large cirque with walls above us. Our pilot swung the plane to the right and I figured he was trying to make room for a hard bank back to the left and crank it around, but he leveled off without turning. I became

scared, glancing back to see if anyone else had noticed our plight. They all had. Gary Johnson had his helmet on and was working with his static line. Sam Houston was throwing on a harness.

We were committed now—no space to turn. The nose pulled up as the engines propelled us, screaming at full power. Outside, the cliffs were coming closer and closer on both sides. Snow and rock. I recall seeing blue ice, a rock cliff, and deep crevasses. Suddenly the ridgetop flashed beneath us. We couldn't have cleared it by more than fifty feet. A wave of relief passed through my body. Rick Blanton shouted, "That was too close, god damn it!" We all were dazed.

Flying along in the Sherpa back to Fairbanks, I couldn't help but wonder how I might feel if one of those near-miss stories had turned out differently. Losing one jumper at a time is disturbing enough. How could we handle a whole planeload? And yet, in fact, we had come close to that very thing in 1986.

A new pilot for one of our Volpars was taking his first check ride dropping jumpers during a practice jump for an eight-man load of rookies. Seeing that the jump spot was near the airfield, he later claimed to have decided to fly with his landing gear down, thus saving himself the trouble of having to activate the hydraulic system twice. More likely, he just forgot to lift his gear.

After dropping four jumpers in two passes, the Volpar was on final for the third with Dan Thompson and Rick Abreau first and second in the door. With the landing gear producing extra drag, the airspeed fell below the stall point. The Volpar suddenly faltered, lurched sideways, flipped over, and began rolling wing over wing. Thompson, pinned by centrifugal force to the side of the door, tried desperately to jump free of the plane but couldn't budge.

"I didn't know what'd happened," recalled spotter Ed Strong. "All of a sudden we were upside down and falling forward on our nose. Everyone in the plane was glued to the side of the fuselage. All this shit they tell us about emergency exits? Forget it! The force was so great I couldn't even lift my hand off the floor."

Dan Thompson remembered it this way: "I knew that if we kept falling we'd just get going faster and I'd never get out, even though my body was just inches from the edge of the door and my right leg was already out. I remember seeing blue sky, dark earth, blue sky, dark earth, spinning faster and faster. I saw visions of a fireball. I was sure we were going to die. I remember thinking, 'I really don't want to die today. Staying with the plane will be certain death. If I jump, will there be enough altitude for my chute to open? Will I miss the propellers and the tail?'"

Terror enveloped him. Then abruptly the force pinning him to the side of the door eased. The plane had fallen into a straight-down dive. Dan put everything he had into one hard pull, cleared the door, slid down the fuselage as the tail flashed past his face mask. Seconds later, his parachute slammed open hard with several line twists.

"When I stopped spinning, I was looking straight at the Volpar. It was close to the ground, fluttering back and forth like a leaf. I felt terrible for the guys on board. Suddenly it appeared to stop, stone still for what seemed like a long time. Then it began moving forward and flew off east toward the Fort."

Dan landed, punching his way through a thick canopy of birch. On the ground all he could do was run around in circles and say over and over rather loudly, "Yep. Yep. Yep. That was pretty scary. That was pretty scary!"

Ken Coe, head rookie trainer, walked over, took one look at Dan, and commented coolly, "Damn rookies. They're getting all the old-salt points this year."

On a wall in the jump shack hangs a photograph taken from the jump spot. In an upside-down dive, with its landing lights on, gear down, and thin vortices trailing from each wingtip, the Volpar spirals earthward. From three thousand feet it fell to five hundred before recovering. Everyone on board was terrified. A few had puked. The pilot landed and was canned shortly after the propellers quit turning. Rick Abreau became temporarily deranged and spent the afternoon circling the jump shack mumbling to himself. Unable to calm his nerves, a few days later he quit smokejumping and never came back.

A week later the Missoula jumpers—many of whom already referred to the Volpar as "the Flying Brick"—came out with a T-shirt showing the Volpar upside down, nose down, just like in the photograph. The caption underneath said: OH WHAT A FEELIN'——DANCIN' ON THE CEILIN'. Needless to say, it was a big seller with all the down-south bases.

Just then the Sherpa slammed into a pocket of clear air turbulence giving everyone a start. It wasn't what we needed right after flying over a downed plane. I tried not to think anymore about crashes and near misses. I needed sleep. Rearranging my gear, I settled down again. But when I shut my eyes, out of the darkness flew *Tanker 138* with its right wing trailing fire.

Pilot Ed Dugan was always telling us about his hair-raising close calls, so the jumpers got to calling him Dead instead of Ed. He seemed to get a kick out of it, and after a while the nickname became so commonplace that it sounded strange when anyone called him anything else.

On a clear July morning in 1981, Dead and one of Hawkins and Powers's C-119s had been taken off duty as an air tanker and refitted for dropping air cargo. Named the Flying Boxcar during the Korean War, the C-119 has two large clamshell doors at the rear. When removed for dropping cargo, they create an opening roughly ten feet high and twelve feet wide and provide a dramatic view out under the C-119's big, thirty-foot-wide twin-boom tail. Once the internal tanks had been removed, roller track was installed on the floor thus making it easier to move the heavy bundles of palletized cargo to the rear of the plane.

The mission that day was to parachute three 500-gallon rollagons—rubber fuel bladders—filled with Jet-A fuel needed to refuel helicopters working a nineteen-thousand-acre fire near Todatonten Lake in the Kanuti National Wildlife Refuge. Dropping heavy air cargo calls for clear presence of mind, teamwork, and good communication between the cargo kickers and pilots. The kickers work at the edge of the ten-by-twelve-foot open door in a deafening roar, exposed to forces extreme enough to pull an airplane in half. Foul-ups can be disastrous. Each rollagon weighs four thousand pounds and is dropped on a one-hundred-foot-diameter parachute. The jumper/kickers on board that day were Tony Pastro, Jim "Oly" Olson, Jack "Fire Pig" Firestone, and "Tuffy" Farinetti.

I remember Oly telling me, "Everything was going good. We were at twelve hundred feet. We'd just sent the first one out the back and were coming around for the second when somebody yelled. Huge red flames were blowing back onto the tail. There was a lot of yelling. We were on fire and loaded with jet fuel."

When "Dead" Dugan saw his right engine explode in flames, he activated the emergency on-board fire extinguisher system. For an instant, the flames went out, but immediately they burst forth again.

"I shut down the right engine and started to feather the prop, but the prop wouldn't feather, and that created a lot of drag on the right side . . . Things got steadily worse after that."

Along with his right engine, Dugan had also lost his electrical system, which disabled the flaps and landing gear. The fire had quickly spread back down the right tail boom and was burning in the tail. He told his copilot, Gary Slocum, to have the jumpers ditch. Slocum rushed down the ladder from the cockpit to the main floor to find the jumpers already gone.

"I didn't need to look twice," Firestone said later. "The fire was spreading down the boom, and there wasn't anything we could do."

Tony Pastro felt the same way. "It all seemed pretty straightforward. You're in a burning airplane, and you have a parachute. I remember when I jumped, I pulled the rip cord pretty quick. One of my shroud lines caught on my leg and gave me a line burn. Suddenly I was out over Alaska watching the airplane I was just in trailing a big plume of black smoke. It didn't seem real. It happened so fast. I remember feeling sorry for the pilots and thinking, they're such neat guys, but they don't have parachutes."

Firestone had grabbed the rip cord of his emergency chute and dived out under the tail. As soon as it opened, he noticed his baseball cap falling away below him. To this day, Jack doesn't know why, but he flew his canopy so that he could watch the cap all the way to the ground. First thing after landing, he ran over and found it. He knew that the proper procedure was to steer your parachute toward the departing plane, join up with the others, and then walk in the direction of the crash. But it hadn't happened that way. Jack wasn't going anywhere without his ball cap.

"We met on the ground," recalled Pastro. "I remember how we were ecstatic one moment for being alive, then sad for Dead and Gary the next. Back and forth—we'd laugh, then think of them, then laugh again. I've never experienced anything quite like it."

Seeing that the jumpers were gone, Slocum had scrambled back up the ladder to the flight deck, where he found Dead dealing with his controls and a decreasing number of options. Dugan yelled for him to go down and cut the restraining straps on the final two rollagons. His plan was to lift the nose high enough to get them to roll out the back. The fire was spreading into the wing, he knew he was going down, and he did not want a thousand gallons of jet fuel tied to his back.

The copilot clambered down the ladder again, grabbed one of the sheathed emergency knives taped to the side of the fuselage, cut the straps, then climbed back up to the cockpit. At this point they had about nine hundred feet of altitude. Dugan dropped the nose, picked up air speed, then reefed back full on the controls, causing the plane to nose up.

The rollagons began rolling down the roller track toward the back door. The plane slowed, faltered near a stall, and Dugan was forced to drop nose. Not quite out the door when the nose fell forward, the rollagons stopped, then came scooting back down the roller track and slammed violently into the bulkhead behind the cockpit. Desperate now, Dugan tried once more. This time the nose stayed high long enough for the two rollagons to travel the full length of the roller track and tumble out.

The copilot grabbed an emergency chute. Dugan yelled for him to get out. Slocum didn't want to leave, but he did as his captain commanded. Rushing down the ladder, he ran the full length of the fuselage and jumped out into fire and smoke. Slocum thought of Dugan as he left the plane, "That's the last time I'll ever see him." At four hundred feet, with the fire eating its way deeper and deeper into the wing, *Tanker 138* flew on.

Dugan stabilized his crippled craft as best he could, all the time searching ahead for a place to put it in. A mile ahead lay the Koyukuk River, a large bend curving in from the left. Along the right side of the river stretched a broad sandbar.

Miraculously, the glide path of the burning C-119 and the extension of the river bar met like two hands reaching for each other. *Tanker 138* settled down lower and lower, then leveled and came in skidding on her belly. She slid about five hundred feet, arcing in the sand slightly to the right, while Dugan held her nose as high as he could to keep it from digging in. In a fury of ripping metal and fire, her right wing caught the ground and twisted her slightly to the right, where she came to a halt crunched forward on her nose.

C-119s are known for their killer noses. When landing on their bellies, their noses often collapse from beneath, buckling up into the cockpit and crushing the pilots. Just as *T-138* came to rest, the floor buckled under Dugan's seat, popping the rivets. Dugan grabbed his camera and flight log from the wall behind him. Then he climbed out on top, jumped off, and ran toward the river. *Tanker 138* was left behind, smoking and creaking. Dugan had his camera ready for the explosion. It never happened. Instead, without the benefit of rushing air, the fire went out on its own.

Such was the last flight of *Tanker 138*. After all her years as a retardant bomber, she had finally died on a cargo run. A helicopter from the fire retrieved the four jumpers and the copilot. All were in good shape. Only Pastro had been shaken up by the line burn and the quick opening of his chute. Slocum said his chute had opened around two hundred feet—two more seconds and he would have died. They flew in the direction they'd seen *138* go down and found Dugan sitting with his crashed plane on the banks of a big sparkling river.

Within the month another T-shirt had been made. This one depicted *Tanker 138* listing right, her propellers twisted in the sand, her back broken, her wings drooping sadly. Her nose had been shoved ungraciously under her chin, crinkling and buckling an already funny face. The old girl appeared quite befuddled at the circumstances of her final resting.

The caption beneath her crumpled nose read: C-119 IN SLOW FLIGHT.

A few miles out of Fairbanks, Togie poked me on the shoulder and said we were preparing to land. Moments later the Sherpa touched down smoothly, taxied the length of the runway, pulled off onto the jumper ramp, and came to a stop in front of the standby shack. As the engines shuddered to a stop, Togie commented to me about the crashed Helio-Courier.

"You almost jumped that, huh?"

"Yeah."

"Boy, that'd ruin your day, wouldn't it," he said, shaking his head. "To be flying along in a plane and have it explode."

14

July 8   Fairbanks

Inside the standby shack I was told by the operations desk that I was to get some rest and be back to work by two the following afternoon. Relieved not to be going to another fire, I went to my locker and found a message taped to it.

Date:.................. July 4, 1992
You were called by:....... Sally
Message:.............. You did it again. Please call.

I told myself to forget Sally and take care of business but then went straight to the phone and called her anyway.

"Hello," someone said. "Captain Bartlett Inn."

"Is Sally there?"

"Sally?... Uh, I don't think so ... Oh, just a minute."

I lingered there feeling like a fool, ass-dragging tired and stinking like a pig. Even my sense of humor was gone. "Dammit!" I mumbled to myself. "Why do you do shit like this?"

"Hello."

"Sally, this is Murry."

"Well, well! So it is."

"I got your note."

"I was just thinking about you. When did you get back?"

"About an hour ago. I just wanted to . . . to give you a call and see how you were."

"I'm fine. Still waitressing. They said you'd be gone awhile."

Sally's voice was like a choir of angels. Images came back. Smiles. Freckles. Curly hair.

"I thought you'd be fishing this week."

"We postponed it. Mike had to work. We're going this weekend."

"It's nice to hear your voice."

"Well, it's nice to hear yours, too."

"I've got tomorrow morning off. Can I see you?"

"Well . . . I could use a ride home. Come by and get me around 10:15."

"I'm a mess. I need a shower. Several actually. And a change of clothes."

"Get your clothes and come. You can shower at the cabin."

Moments later I was in my pickup, turning up the music, singing with it, and heading for the barracks.

After I picked her up, we stopped for a bottle of wine and a pizza, then drove to her place, where I showered until the hot water ran out. Sally changed into faded Levi's and a tank top, lit a candle, placed it on the coffee table, and poured two glasses of red wine.

"Here's to Sally and all the fish she's going to catch," I said, lifting my glass to hers."

"And here's to me, too," she said, obviously pleased.

Candlelight splintered richly in the red depths of the wine as we sat there quietly enjoying the sweetness of our surprise meeting.

We talked a little longer, put away the dishes, kissed in the kitchen for a few minutes, then went upstairs to the room with the big green quilt. I stood at the window looking down into the birch woods as the midnight sun turned wildwood roses into pink butterflies. Cottonwood seeds drifted lazily in a golden haze, lifting and falling effortlessly, then disappearing in shadows.

Sally came to the window to look. We kissed, feeling the silky air flow around us and into the room. I lifted the tank top straight up over her head and dropped it to the floor. Sally unbuttoned my shirt, and we stood bare from the waist up, tangle-footed, chest to chest, kissing. I tugged her Levi's off over the firm hump of her buttocks and let them slip down, exposing her legs in a wash of pink light. I fondled her for a time until she stepped clear of her jeans. Sally then did the same to me until we were standing naked in full embrace, tugging and pulling at each other. For a time that was enough. Then the bed, the green quilt, and Sally's red hair rocking back and forth above me in the silence of an approaching dawn.

July 9

Scanning the jump list, I quickly assessed the fire situation statewide. Of the 235 smokejumpers in Alaska that afternoon, approximately 190 were on fires, 37 were back in Fairbanks, and 8 were listed under "cripples," including Al Seiler.

"How close are you to being fire ready?" Battaglia asked me. George had taken over the operations desk as the new box boy.

"Half an hour."

"Sounds good. You're the bottom of the third load, but I wouldn't waste any time. Things have been crazy around here every afternoon. We've dropped an average of sixty jumpers a day for the last twenty days."

"How's Al doing?" I asked.

"Seiler? Not that great. The brain scan came out OK, but his eye's a mess. The impact from hitting the tree may have torn loose some retinal nerves in the same eye the sliver's in. *Not* good news. He may lose it. The doctor doesn't know for sure. He's inclined to operate on the sliver, but I guess he took quite a bit of time explaining the risks of cutting that deep. The only other alternative is heavy doses of antibiotics and waiting to see if the sliver will dissolve. If the eye gets infected, he'll lose it for sure. Al's taking a couple days to think it over."

I went to my locker and opened it to a mound of trashed-out gear. My PG bag sat crusted with dried fire retardant and stuffed with dirty socks, T-shirts, and partly eaten plastic bags of jerky and dried fruit. My jump boots were caked with ash and mud as hard as concrete. One sleeve hung in tatters off my fire shirt. Rolled up in an ugly ball of canvas and tangled lines was my tent and tarp.

Glancing around, I noticed a photo taped to the front of Seiler's locker. Kneeling in full line gear, Al and Rene Romero are intensely involved in assessing a problem the first night of the Clear fire. Shoulder to shoulder, heads tipped inward, they sight down the length of Rene's right arm in a posture of complete unity of purpose. And although Al's eyes burn clear and hot as he listens to Rene, I am struck by the humility and goodwill I see there. Here is a man whose heart lives right behind his eyes.

I sank down next to my locker, dropped my face into my hands, and found myself fighting a sudden and overwhelming sense of despair. Tears welled up in my eyes. Al was hurt. So were some others. I shook my head to clear it. What would they do? I got to my feet.

Angry, but determined, I dumped my gear bag out on the floor, separated my jump gear and began the task of refitting. I stuffed my gear bag inside my jump jacket, my tent inside the jump pants to cushion my butt, then refilled the leg pockets with my letdown rope, tent poles, long underwear, baseball cap, and radio. Refilling my canteens, I completed repacking my PG bag, put a new main on my harness, and hung the whole outfit on the speed racks. With my gear squared away, I yelled to Battaglia that I was fire ready and headed to the loft to help pack chutes.

"Hey, now. Look who's decided to come to work," Charlie Brown announced as I walked in. "The Love Machine. Complete with a terminal case of moose eyes."

"Old Leathersack. Feeling a little moosey, there?" asked Tyler Robinson.

I ignored them and walked straight ahead, until I reached the square tower.

"Go ahead, hide!" Charlie carped. "Right in the middle of the bust . . . The old man gets moose-eyes for some little redhead, takes the morning off, and leaves his bros to pick up all the slack. Here's a real professional in action. Runs out on his buddies when the going gets tough, and for what?"

In that instant, the siren sounded and we all started running to the ready room. At the speed racks we helped the first load suit up.

"It's down toward Denali," Battaglia shouted. "Another railroad fire—right next to the mountains."

As the first load cleared the ready room, I headed for the rest room to make sure I wouldn't get caught on a long flight with a ballooning bladder. My position on the jump list had moved up to the bottom of the second load.

Five minutes later the siren blew again. The call was for the same area. Another railroad fire, one more load. That made me last man, first load. Moose-eyed or not, all my hopes of spending another night with Sally went out the door with that second load. I went to the speed rack and rechecked my gear. The siren went off again.

"To the north," Battaglia shouted. "Toward Fort Yukon."

A note came from the cockpit and was passed back to the first man, Jim Kitchen. The fire was ninety miles from Fairbanks, and a helicopter crew from Fort Yukon was also responding. There were no specifics. Since it wasn't far, we did our buddy checks early. Approaching the Yukon Flats, we passed a large let-burn fire off our left wing, burning along peacefully all by itself. Located in the Yukon Flats National Wildlife Refuge at the east end of the White Mountains, it looked to be about ten thousand acres.

Farther out in the flats we found two small fires. The first had been taken by the crew on the helicopter; we dropped two jumpers on the other.

We circled many times, making low passes to find a clearing for a jump spot. An old burn, the area was a sea of dead trees all bristling with sharp, iron-hard tips. For Togie and Kubichek all the circling turned out to be too much. They turned white, then green, then pulled out their plastic bags. They managed to keep their misery to themselves, pinching off the bags tightly to keep the smell down—smokejumper etiquette. It was fortunate that they did because the flight was rough and the domino effect threatened the entire load.

Finding a safe jump spot is always taken seriously, especially since Gary Dunning's July 4, 1986, mishap, which happened not far from where we were flying.

Gary was drifting under a round parachute in a strong wind. He knew the penalties of turning and running with the wind, for that would have given him a ground speed of thirty to thirty-five miles per hour. Even holding against the wind, he drifted backward across the ground at ten to fifteen. Gary held his position, facing into the wind, not seeing where he was going, just hoping for the best.

Gary jumped most of his twenty-five-year career out of West Yellowstone but also spent several seasons in Cave Junction and Alaska. As he told me his story, he paused and looked around warily, jostling his stocky frame like a boxer ready to duck punches.

"Next thing I know, I hit this damn tree and hang up. After all the crashing stopped, I'm sitting there looking out over the tops of the other trees. I started trying to sort out the mess so I could get on with my letdown. Something was wrong, though. I couldn't tell what, but you know . . . something.

"I flipped up my face mask so I could see a little better. Once I unsnapped my reserve, I saw it . . . It was kind of unreal. Then the pain hit."

Drifting backward, Gary had come down on top of a black spruce snag. Its tip had pierced the Nomex of his jump pants and punctured through his right thigh. Gary was holding on to the top of a tree with its tip run through the middle of his thigh a full four feet.

"I couldn't believe what I was seeing. You know we never figured stobs could puncture Nomex. I didn't really panic. It wasn't believable enough to panic, but then I got to thinking that I must have cut the femoral artery. If that was true, I knew I'd bleed to death if I didn't get help quick. There didn't seem to be a lot of blood, but then again I couldn't see it very well. I must have been up there for five minutes, and I kept thinking that the guys would come over, but I didn't know if they knew where I was. I yelled and yelled, but the wind, you know. Pretty soon I began to feel kind of shocky. All the time the wind kept pulling on the parachute so I cut it loose. Then I started rocking back and forth as hard as I could, and finally the trunk snapped off about ten feet below me. "

The top fell thirty feet to the ground with Gary skewered through like a piece of shish kebab.

"I was knocked cold when I hit. Next thing I know I'm coming to and feeling pretty woozy. I could feel blood running in my pants and down my leg. Some guys came over, saw what had happened, and ran to get a chain saw. They cut the tree off on both sides of my leg and just left it in there. That's the way they brought me in."

Smokejumper physician Dr. Carey Keller treated Gary when he got to Fairbanks. On his office desk he keeps a photograph of Gary with the stob still in his leg. Remarkably, the stob had found its way precisely between two major muscle groups, barely missing the femoral artery.

As we climbed to jump altitude, spotter Buck Nelson gave his briefing to the two jumpers in the door. Throughout the old burn, streaks of red and purple fireweed ran wild into miles of rolling lavender hills.

Kitchen and his jump partner readied themselves in the door as Buck took his last look under the plane. Then he lifted his arm and they braced. Down came the slap on Kitchen's shoulder, and off they went. Immediately the Dornier poured on the power and started into its turn. We looked back and saw them there, tiny figures hanging under their parachutes, sailing above that beautiful country. To the east lay the Yukon River and the massive chalk bluffs forty miles upstream from Circle, while off to the north stretched the magnificent Yukon Flats, a great slab of green under a dome of blue and white sky.

As soon as Kitchen and his jump partner got on the ground, we picked them up on the intercom inside the jump ship.

"Yeah, yeah," Kitchen sang out. "We're OK—no problem!" His voice trailed off in laughter. Then he was back again. "You bet we're OK! How you guys doin'?" Laughing again. "What a day. We're OK. We're *way* OK."

We dropped their cargo and pulled up for Fairbanks. Togie and Kubi looked near death. Smiling lamely, they passed their plastic bags back to Buck. He eyed the bags as one might dead skunks, then accepted them at arm's length and tossed them quickly out the door.

When we waddled back into the ready room it was 6:10 P.M. Hot, half sick, and soaked with sweat, we unsuited, and hung our gear back on the suit-up racks. While we'd been out flying, the load behind us had rolled to a fire.

"Man," the box boy said. "You guys look rough. Feel like eating?"

All except Togie decided to eat. Naturally, Kubi was into it, hunger being a more or less permanent condition with him.

Back at the shack things remained calm. We had fifteen available jumpers in Fairbanks. Two loads had made it back in to Galena. Rain was falling in the local hills; lightning in the interior had died down. Still, I was certain there was little chance of spending another night with Sally.

Resigned to our fate, the fifteen of us hung out in the lounge of the standby shack watching TV, thumbing through magazines, and falling asleep in chairs and on the floor. Nine o'clock came and went. No air tankers took off. No helicopters beat the air around the main helibase. Everyone was edgy. I had my fingers crossed—I think we all did. At 9:50 Battaglia came on the PA system. "Everyone off at ten."

When the second hand hit ten, I was out the door and never looked back.

"Be here at eight o'clock sharp in the morning," a voice from the box shrieked as the door slammed behind me.

It was 10:30 when Sally came out the back door of the Captain Bartlett. We stopped for a six-pack, drove out the Parks Highway to Sheep Creek Road, and then turned off on the road that led to the top of Ester Dome. After almost two months, Sally had seen little more of Alaska than a few back roads around Fairbanks. With that in mind, I drove as far as I could out past the telecom towers and parked on a high overlook facing southwest, where we could see the Tanana Valley, the long hills that run west to Nenana, and far beyond out into the interior. Through air cleansed by afternoon rain, the Alaska Range appeared as a jagged white wall etched in blues, culminating in Mount McKinley's massive twenty-thousand-foot summit. Twenty miles to the south, the North Pole oil refinery glittered brightly in reflected sunlight.

"It's the Emerald City," Sally called out. "The yellow brick road must be here somewhere."

"North Pole refinery," I said. "It's on the pipeline from Prudhoe Bay to Valdez."

"It's not, either. It's the Emerald City, and I'm the cowardly lion. Put 'em up, put 'em up, put 'em up." Sally snarled fiercely, fists clenched.

"I'll fight you with one hand behind my back." She thrust one hand behind her back.

"Put 'em up, put 'em up," Sally demanded. "I'll fight you with *both* hands behind my back."

Sally reached for my beer, downed the last of it, then handed it back to me smiling. "I like you," she said. "I hope we can do some more things together. Maybe, someday, even go canoeing."

I cocked my arm out the window and tossed the empty can into the back of the pickup. As I did she nuzzled me on the side of my neck, and we began kissing. After a few minutes we broke away, cooled down, then started in all over again.

After a half hour of high school–style smooching, I got out one side of the pickup while Sally got out the other. Walking a ways from the pickup, I stepped behind a bush. Sally went the other way and took a little privacy of her own. I picked a couple sprigs of fireweed, some yarrow, a few wildwood roses, and bunched them up into a bouquet. What Sally had said about more time together was a dream come true. Soon fire season would slow down, and the summer would be ours.

When I got back to the pickup Sally had opened another beer and was handing it to me, so I handed her the flowers in exchange. Bouncing up and down on the edge of the seat, she waved them under her nose, shut her eyes, and inhaled. I pulled her curly red head down under my chin. Blindly she held the flowers up for me to smell, ramming them in my face. Laughing, we rocked back and forth, holding tightly.

Knowing about her degree in art, I asked, "Been doing any drawings?"

"Not since school. I don't have any stuff here—just a few sketch pads, pencils, and chalk."

"Will you make me a sketch?"

"Of what?"

"I don't know. Something that'll remind me of you."

The sun slipped below the rim of the earth, and the Emerald City went dim while the upper Tanana Valley settled into an indistinct haze.

Hearing an airplane, I looked up to see the Dornier, bright blue and white, its nose held high, climbing west.

"That's one of our planes," I said. "The one I flew in today."

"What's that mean?"

"Either they're rolling to a fire, or they're on a paracargo run."

We watched the plane climb high and disappear beyond the clouds.

"It could be going to bring back jumpers from Galena, but most likely it's a fire call. The Dornier was the first-load ship when I left."

Sally sat smelling the flowers and staring straight ahead.

"That'll put me number one first thing tomorrow."

"You don't know whether you're coming or going, do you? You could go tomorrow and not be back for a month, and you don't even seem to care."

"I care," I said. "But right now there's a limit to what I can care about."

Sally was quiet a moment.

"It's dangerous, isn't it?"

"Sometimes," I said. "But mostly not. We know our limits. When you choose to place yourself in danger, you learn to live with it."

"Then you're not afraid?"

"It's not that. There are just times when I have to perform, afraid or not. Al Seiler told me about circling a hot fire, crowning in big trees on a steep hillside. It was his first fire jump. The spotter asked him 'Are you scared?' He thought about it a second and said 'Yes, I am.' 'That's OK,' the spotter told him. 'We're all scared. Get in the door.'"

Sally and I drove the winding road down Ester Dome and headed back for Four Corners, she sitting next to me holding the flowers in her right hand and resting the left on my thigh. When we pulled into the yard, a pang of apprehension shot through me. Most likely I'd not be welcome overnight if her brother was there. Thankfully, his truck wasn't in the garage. For a moment the two of us sat quietly in the front seat of the pickup.

"You know, Sally," I said. "Speaking of my job. A few years ago an entomologist from the university showed up in the ready room at roll call one morning. This guy had spent his whole life collecting rare butterflies. He asked if we'd catch butterflies for him and bring them back to Fairbanks. He told us that we were going to places that few people had ever been and were probably seeing butterflies that no one else had ever seen before."

July 13   Noorvik

Three days ago we jumped the Kobuk fire atop a windy bluff above the Kobuk River. Early this morning, while the rest of us ate breakfast, the fire boss flagged down a boat and caught a ride downriver to Noorvik to telephone Galena dispatch. Our eighty-acre tundra fire was out, except for a few smokes that persisted along a big slough between the fire and the village. As the rest waited in camp for his return, I took a walk east along the crest of the bluff toward the Kobuk Valley National Park.

A great dome of sky arched over a treeless landscape of creamy greens and yellows. It was clear and windy. A hundred miles to the northwest stood the mountains of Noatak Natural Preserve and east of them, toward the top of the world, the Brooks Range. To the west lay Hotham Inlet, Cape Krusenstern National Monument, and a hundred miles beyond that, the Chukchi Sea, where the curvature of the earth described a thin gray line under a veil of orange smoke drifting out of Siberia.

During the Pleistocene epoch, 12,500 years ago, this valley had been the eastern extension of the ice-free corridor that crossed the Bering land bridge between the Eurasian and North American continents. Looking out over the sea, I tried to imagine the land that now lay beneath it—a passageway for hunter and hunted alike, a vast savanna, home to herds of giant herbivores, saber-toothed tigers, and some of North America's first people, their bones now buried under the sea. Migrating across the land bridge, up the Kobuk Valley, and on into the interior, they left behind some of the oldest spear points, bone knives, and stone tools found in the region.

During our stay on the fire, several boats from Noorvik had stopped. Their passengers had been traveling the river, keeping an eye on the bluffs. When the river runs high, silty soils undercut and give way. As the banks collapse, fossilized tusks of bygone woolly mammoths come curling out into the light and air, exposed once more to man the hunter. The tusks are cut and carved and sold to the ivory traders.

Looking back toward camp, I took a moment to survey the bluff and the black fire scar that ran across the broad plain. Our blue and gray tents sat at the edge of the bluff, buffeted by the wind. On our first day, Togie had found a set of sun-bleached caribou horns—skull and all. He'd packed them to camp and mounted them on a pole at the head of my tent. For three days of constant wind, the horns restlessly jerked back and forth, the spirit of the animal still with us, even in death.

Walking back along the bluff, I watched as a boat trailing a grace-ful V upriver slowed and put ashore. Bill Cramer, our fire boss, climbed out onto the muddy bank and began negotiating the arduous trail up to camp.

"Well," Cramer said, his face beaming. "Are you guys ready for the latest?"

Ray Brown growled at the young jumper as might an old bear at a cub. I groaned and looked off out over the river. Tired and wornout, everyone else kept staring into the campfire. Most of the time smoke-jumpers are not only ready for "the latest," but starved for it. Hour upon hour is spent speculating upon speculation, only to speculate fur-ther in hope of providing some measure of relief from the quandary of not knowing what will happen next.

"Latest what?" Ray grumbled into his third can of pork and beans. "The latest cretin dispatcher trying to ruin our fun?"

"They want eight of our fourteen to demobe immediately by boat to Noorvik," Cramer said with authority. "A plane's coming to pick you up and take you back to Galena in a few hours."

A collective moan went up around the campfire.

"Just when I'm about to get in some fishing," Ray said. "Screw Galena. They don't know shit anyway."

"Apparently," Cramer went on, "they want all the jumpers they can get into Galena before nine tomorrow morning. From there they'll be picked up by Evergreen's 737 from Boise and flown directly to Fairbanks."

"What the . . . ?" Ray croaked, bucking forward and snorting a few pork and beans out his nose. "Are you out of your mind?"

"No shit!" Cramer said. "The Lower 48 got hit with lightning. They're at preparedness level II with red flag warnings in the North-west—fifty jumpers are going south tomorrow."

Eight of us rushed to strike our tents and haul our gear down to the river. At one o'clock, a twenty-foot flat-bottomed boat from Noorvik arrived and we said our good-byes to the six that would remain behind. As the boat drifted from shore into the swirling current, I looked up and waved farewell.

Our boat landed on the broad sandy beach at Noorvik, where a dozen brightly painted fishing boats had been dragged to higher ground, lined up side by side, and turned upside down. Draped over wooden fish racks, green plastic netting hung drying in the sun. The Kobuk River is a clear river, blue-green and cold. As we landed, its surface reflected brightly on the riverbank, the bluffs, the old weathered buildings, and the painted boats.

Located thirty miles up the Kobuk River from its mouth at the east end of Kotzebue Sound, Noorvik is an Eskimo village of about 550 people on the edge of a broad tidal flat where the land slowly and imperceptibly gives way to the sea.

An older man, the father, I think, of the man who'd picked us up, came down and generously offered to take our heavy gear to the airstrip in his four-wheeler and trailer. He laughed a wonderful silent laugh. We thanked him repeatedly because we each had about 125 pounds, counting the extra parachutes we were expediting back to Fairbanks.

"Which way's the store?" Dalan asked.

"Main Street," the old man said, pointing to a cluster of buildings above the beach.

We went down to the water's edge and scrubbed our hands and faces with black sand, then headed for the store.

You could have filled a shopping cart with all the junk food we bought: bags of peanuts, ice-cream bars, licorice whips, candy bars, cans of Pepsi. Across the road we took our seats on discarded fifty-five-gallon drums and ate and drank. Tires popping on gravel alerted us to three native boys riding up on bikes. They came to a stop not six feet away. We grinned, they grinned, seven- or eight-year-old Dennis the Menace types—scruffy looking, mischief incarnate. The biggest one hauled his little brother along, squatting in a wire basket behind the seat. These youngsters, at least initially, appeared to approve of us. We obviously shared their love of junk food, and besides, we were old enough to keep ourselves clean but apparently preferred not to.

"Hi there," Dalan Romero said. "And just what are you boys up to?"

Trading glances, they rolled eyes, tightened lips, suppressed urges to explode in laughter. Apparently no question could have been more ridiculous. Giggling, they stood straddling their bikes, cranking the front wheels back and forth while casting dubious eyes at everything we apparently stood for. They blurted a few quick comments in Inupiat, then peddled away howling their heads off.

A radio was playing in an open window of the store, and we listened as the music ended. The station was broadcasting from Kotzebue, and the announcer's English was inflected with the endearing, cryptic accent of the Eskimo.

"I got new album here," he said, " . . . but, ahhh . . . new album. Don't think you heard it yet . . . new album by . . . Chuck Berry. But first, I got message from city council: 'All you people who got caribou skins drying in your front yard, take them away. They don't look good . . . and they stink . . .' And now . . . Chuck Berry."

On our walk from the store to the airstrip we came first to the oldest part of the village, which had been established around the turn of the century. We could see where the oldest part ended in a line of broken-down log structures, bleached gray and abandoned, but could only guess that the village had had its beginnings nearer the river and had retreated inland as the bluff had eroded. From old to new, the rough log homes marked a steady progression forward in time, the remains of the oldest cabins having completely fallen in on themselves, now looking like old ships sunk into a sea of bronze grass.

Farther up the road is the newer part of Noorvik, the part that was populated from about 1930 to 1975, between the time when white men arrived in numbers and the construction of the most modern part of Noorvik, called New Town. Russian explorers arrived here in the early 1800s, followed by gold seekers stampeding up the Kobuk River in the wake of the Nome strike of 1898. But it wasn't until the increase in fur trading in the 1930s that the northwest coastal villages of Alaska were significantly impacted by the outside world.

Those times were reflected in this second part of the village, with its milled lumber, metal roofing, glass windows, metal stovepipes, rusted heaps of vintage snowmobiles, and old pickup trucks. Most of it, too, had been reclaimed by the waist-deep grass that flowed before the wind, revealing, hiding, and revealing again a past settled among purple fireweed.

New Town is where most of the people live now, on straight streets, in uniformly constructed frame houses built for the most part with money that came after the Alaska Native Claims Settlement Act of 1972, the same legislation that settled once and for all who would hold title to what lands in Alaska. Besides the eighty million acres of national parks, monuments, forests, and wildlife refuges it created, the act provided for native Alaskans to claim individual parcels and paved the way for the designation of the larger allotments, now managed by the various native corporations.

The act changed forever the status and much of the structure of native societies; it cleared the legal right-of-way for the building of the Trans-Alaska Pipeline, only the first of many big-scale projects envisioned by development-minded Alaskans who are effecting the wholesale division, subdivision, patenting, parceling, and deeding out of physiographic Alaska.

Between the houses of New Town, mini–wrecking yards contained dozens of items old and new. There were snow machines of every possible description, from the bright shiny new to the gutted-out wrecks that had been cannibalized for parts. There were three-wheelers, four-wheelers, motorbikes, and no-wheelers. There were outboard motors, boxes, tarps, flatboats, fish racks, rafts, sawhorses, spare tires, stove parts, cans, buckets, tires, wheels, woodpiles, fishnets, and piles of caribou hides and horns, and the ubiquitous fifty-five-gallon drums.

On a front porch a family sat visiting with neighbors, smoking cigarettes, drinking coffee, laughing. Off to the east lay the broad tidal flats—the eternal flats, bright with a wild and clean slant of light. The flats symbolize a people's patient waiting for summer's brief passage and the return of their beloved winter. Native Alaskans feel more alive, more active, more healthy in winter. They get out and move in winter. The rivers, sloughs, lakes, and the sea all freeze tight and are easily traveled. Winter is freedom. Freedom to roam their wilderness world locked in solitude and isolation from the outside. Winter—when the sun barely clears the horizon before it's swallowed up again by the long Arctic night.

Continuing our trek out to the airstrip, we left Noorvik behind, clinging to its eroding bluff, an island of humanity nestled amid the rubble of its past.

While we waited for our plane, sprawled out in the sun, the old man who had helped us haul our gear came walking out the road from town and stopped beside us. He was wearing clean khaki trousers and a black-and-white checkered cotton shirt.

"We were thinking the fire might burn the cemetery," he said in a voice giving the cemetery its due respect.

Those of us still awake lifted up on our elbows.

Thick legs filled the old fellow's trousers. His shoulders were powerful looking, angular and full. His hair was pure white.

"It's not going anywhere now," Dalan assured him. "It had to jump a pretty big slough before it could have burned into town."

"Yes," the man agreed. "It's a pretty big slough."

The old man told us of blizzards he'd seen when the snow got so deep that houses were buried and people had to dig tunnels down to their front doors and shovel their stovepipes clear so it didn't "smoke them crazy."

"Used to trap with my father when I was boy . . . up to the Noatak."

"How long would you stay out?" Dalan asked.

"Four, five weeks sometimes."

"With a dogsled?"

"Dogsled and snowshoes."

"How much distance did you cover?"

"I don't know. Maybe five, six hundred miles."

"Did you have cabins?"

"Two cabins," he said, his face suddenly merry. "Small ones . . . but mostly just a tent."

"Just you and your father?"

"Dogs, too. All winter trapping with my father, up to the Noatak."

Now a natural preserve, the Noatak Valley is a beautiful empty land following the river north and east two hundred miles into the west end of the Brooks Range.

"When was that?" I asked.

"Oh," he said, laughing his silent laugh. "Long time ago. I was just boy."

"What kind of food did you take?"

"Dried fish skins, smoked fish. Seal oil, rice, sugar, tea, dried fruit, moose, caribou, sometimes nuts."

"The way we live now is cockeyed," he told us flatly, looking off to the low hills. "We don't follow caribou no more. Used to follow caribou and hunt them, some years upriver, some years out to Noatak . . . other times north of Kiana hills. Now there's hunting season . . . hunting license." The old man hesitated and shook his head. "It's cockeyed now."

He looked down the runway as if he heard something. We looked with him. Nothing. Then he looked toward the hills again.

"The weather's cockeyed, too. No weather like this when I was boy. People not teaching old ways. Other things . . . the animals know it, too."

He kept his eyes on the horizon to the northeast while the wind flicked the fine white hair around his ears. At last he sighed deeply.

"Did you hear what happened up to Kobuk?"

No, we said. We hadn't. The old man paused at length.

"Yesterday . . . yesterday morning, up to Kobuk. Sled dog broke his chain, killed a little two-year-old girl. Ate most of her before they found it."

The old man stood there, his eyes calm and far away. A gust of wind rattled loose metal on a maintenance shed and rushed toward us in a small dust cloud. Without another word, the old fellow turned and walked away. We muttered a feeble good-bye to his departing figure. He walked the single-lane dirt road toward the village, growing smaller and smaller, and was soon gone.

15

Morning came early. For a moment I lay between clean sheets, reveling in the luxury of resting my head on a real pillow. My knee had a sharp pain right below the kneecap, along with its usual morning stiffness. A soft stirring traveled up and down the hall of the Galena barracks. I got up and put on clothes that smelled rancid with sweat and smoke, then took a look at myself in the mirror. Matted hair crowned a weather-beaten face that stared back at itself with lemur-like eyes, large, bloodshot, uncomprehending. I rinsed my head under the cold tap, slicked back my hair with my fingers, pulled on my baseball cap, grabbed my PG bag, and stepped out into the hall. *Always leaving,* I thought. Trying my best to follow the crazy, haphazard script.

Walking into a warm cookhouse filled with the smells of fried bacon and homemade cinnamon rolls is a sensual experience after spending three days on a cold, windy bluff eating pork and beans and smoke. The dining hall was clean and orderly. The floor was flat, and there were good places to sit where you didn't have to hold your food on your lap. Clean white cups sat in neat rows, waiting to be filled with steamy black coffee. Bins overflowed with bacon, sausage, and ham. There were fried potatoes and eggs, stacks of toast, mounds of pancakes. One table was crowded with trays of cantaloupe, green melon, apple slices, pineapple chunks, grapes, and cherries. Another held glasses of orange, grape, and apple juices, bowls of granola, yogurt, and mixed nuts, along with fresh homemade biscuits.

Take fifty famished smokejumpers, fifteen Galena support personnel, four air tanker pilots, various mechanics and fuelers, and add thirty other fire-fighting types clomping around in heavy boots and you have the Galena chow hall at 0630, July 14, 1991. And, of course, we can never forget the cooks themselves, Allyson, Holly, and Eileen. To the hungry, the battered, and the sometimes lonely, these dear women appear the loveliest of angels, friendly, clever, bright-eyed, and beautiful.

These three ladies took their task of feeding hungry firefighters every bit as seriously as we took packing parachutes. Taking a moment to chat with them, I noticed a collection of cookbooks sitting on a nearby shelf: *Jane Brody's Good Food Book, Cheesecake Madness, Classic Indian Cooking* by Julie Sahni, Madhur Jaffrey's *Far Eastern Cookery, The Best of Gourmet.*

At seven o'clock Dalan Romero came in from dispatch and passed the word around from table to table.

"All jumpers be on the hard ramp at 7:15. The 737 is ten minutes out. Fairbanks is crying for jumpers. Last night central Idaho and eastern Washington got pounded by dry lightning."

We finished eating, gave profuse thanks to the cooks, then filed out into a calm, red and gray morning just as the 737 touched down.

Within the hour we'd landed at Fairbanks and were back in the standby shack. Roll call was just getting under way. During our absence a new box boy had barnacled onto the ops desk.

"OK, listen up," Al Seiler shouted, attempting to settle the general tumult of the ready room. A black patch covered his left eye. The removal of the splinter had apparently gone well. The nerve damage wasn't as serious as first thought. In the meantime, the base manager had placed Al on light duty.

"The Lower 48 wants fifty jumpers south today," Al told us. "Forest Service boosters will be first, starting at the top of the list. If we don't get enough that way, then it'll be choice from the top of the Alaska list. If we still don't get enough, it'll be boned from the bottom. I don't think a whole lot's happening yet, but things are heating up. Preparedness level II, Red flag warnings. Ha, ha . . . and here's me stuck looking like a pirate. All I need is a peg leg and a parrot, and I could go as an anorexic Long John Silver."

Al went down the jump list, calling off names and sliding name tags to the side. The fifty needed were available in the first round of boosters. With them off the list, that left me somewhere down on the fifth or sixth load.

I drove to the barracks, took a long shower, got into some clean Levi's, and spent the remainder of the day organizing my jump gear, restocking my PG bag, and mulling over what appeared to be a break in the action. In the training room I stocked up on ibuprofen and stepped on the scales. In two months I'd eaten everything I could get my hands on and had still lost ten pounds, down now to 166.

In the loft, jumpers were quietly packing chutes, mending gear, and listening to Kenny Coe's Gene Autry tapes. I felt a pang of envy for Gene, a man who apparently finds contentment drifting with tumbleweeds and singing to a horse.

In the lounge, bodies were sprawled across couches, slumped in chairs, and stretched out asleep on the floor. In the aisles between lockers, small groups of jumpers sat and talked. A few wrote letters. I sorted my mail and kept trying to call Sally but couldn't reach her.

We rolled two loads that afternoon, not to fires, but to pre-position them in Fort Yukon and Palmer. Two loads were held on standby until eight that evening. Not a prop turned.

July 15

Alaska fire activity had slowed down—typical for mid-July. As of midnight the fourteenth, 786 fires had charred 855,673 acres. Smokejumpers had made 1,080 jumps to 146 fires, an average of 7.4 jumpers on each. Nearly 190 fires were being allowed to burn in the limited areas. Four major fires (one still out of control near the village of Anvik) made up 50 percent of the total acreage. Besides the 65 smoke-jumpers on the regular Alaska crew, 170 extras had been called up from the Lower 48—fifty of those left for Idaho yesterday. In eighty-three separate paracargo missions, 245,828 pounds of fire line supplies were delivered via fixed-wing aircraft and parachute.

During the height of the bust, the Alaska Fire Service had ten medium and five light helicopters on contract. The State of Alaska had three heavies (U.S. Army), eleven mediums, and eight lights flying for them. Those we shared when possible. The thirty-seven helicopters had worked hundreds of fires, flying an average twenty thousand miles per day for five weeks for a total of seven hundred thousand miles, or a distance just short of two round-trips to the moon.

Although federal and state fire organizations continued working in the extended attack and support phases of the various fires, for smoke-jumpers the action had slowed considerably. Basically we were back in position, regrouping, and making preparations for the next go-around.

In the shack, fire reports were scattered all over the operations desk; time slips heaped in alphabetical stacks. Parachutes began filling the empty slots in the storage bins, and the backlog of unchecked and unrigged parachutes was decreasing steadily. I spent the day filling out overdue reports, checking chutes, and talking with our physical thera-pist about my knee.

For the past three weeks, I'd taken ibuprofen daily trying to keep the swelling down. The pain cycled between a constant dull ache and the occasional ice-pick stab that caused me to cry out involuntarily. The bros, of course, began imitating my wails, apparently finding them a rich source of entertainment.

Our therapist was a tall fellow in his early thirties with a long face, big feet, and closely spaced eyes that sparkled under a high mound of dark curls. He listened intently, sitting on his stool, hands folded in his lap, his eyes watching my every move. By nature a reserved young man, Jim Kimbal spoke softly and never without forethought.

"How long has it been bothering you?"

"I first noticed it in Montana in 1988. Every year it gets a little worse."

Jim had me lie down on a table, where he examined both knees, flexing each leg one way and then the other, his ear held close, listening for grinding sounds. On the knee that had been hurting, he repeated one movement several times, listening carefully.

"Overall," he finally said, "your knees sound pretty quiet. There's some grinding in the left one, which indicates calcium buildup on the kneecap . . . the beginning stage of arthritis."

He flexed the injured knee again, pressing with his fingers on different points. I jerked when he hit one spot. Placing the leg back down on the table, he sat back on his stool, folded his hands in his lap, and smiled.

"Based on your age and how long you've been jumping, I'm guessing your knees are pretty beat up. It might be time for an MRI. On the other hand, both knees seem structurally sound. The muscles supporting them are well defined and doing their job. The anterior cruciate ligament of your sore knee's been hyperextended but not enough to need surgery."

He hesitated, his face expressionless, thinking about what he was saying.

"The number one source of the discomfort in your left knee is probably the meniscus. The meniscus is a pair of crescent-shaped cartilage pads that cushion the ends of your leg bones. My guess is that yours have multiple tears along the inside edges. Once they start, they're like the cracks in a windshield. They grow and grow until they begin irritating nerves."

The meniscus, I thought. It figured. One of every five on our crew had been operated on for a torn meniscus.

"On a scale of one to ten, how bad is it?" I asked.

Kimbal smiled. "Oh . . . it's a ten. I'd cut out the running for now. Concentrate on strengthening the muscles around your knee. Use the stair-stepper, the stationary bicycle . . . also, leg extensions are good. Take it easy for a while. If it hurts to exercise, hold off. Strengthening the muscles will also protect the anterior cruciate ligament from further hyperextension. Keep taking the ibuprofen for now. Watch your dosage, though. Too much can hurt your liver. Meniscus damage doesn't heal. When the pain gets bad enough, you'll need arthroscopic surgery.

"Jumping's beginning to take its toll on your knees. I know you like the work, but too much of this kind of damage could cause you problems later on."

"Like what kind of problems?"

"Like problems walking . . . depending on the type and the seriousness of the arthritis and how your body reacts to it."

"It doesn't seem that bad. Except for the pain, I feel good. But I hear what you're saying, Jim. I don't like it, but I guess my time's coming."

"Other than your knees, you're in terrific shape. Still, you don't want a disabling injury this late in your career."

"Thanks, Jim," I said, shaking his hand, standing up to leave.

On the face of it, I felt relieved. The news could have been worse. Barring specific injuries to other body parts, a smokejumper's knees are the first thing to go. I'd been lucky mine had lasted so long.

July 17

Summit Lake lay brooding between black hills, the steely chop of her waters shimmering before an approaching storm. My windshield wipers flopped back and forth in a pelting rain as the pickup cleared the summit and began the long descent toward Glen Allen and the Copper River country. On the road for four hours, I estimated at least that much more to go. I didn't know for sure; I'd never been to McCarthy before.

Things had gone dead at the shack. Except for those Lower-48 jumpers still out on fires, most of them had been sent home. In the aftermath of so much action, an odd restlessness had settled over the crew. We had plenty of free time, but no one seemed to know what to do with it. Having not been able to reach Sally had only added to my anxiety. Then, yesterday afternoon, a man answered the phone in a gruff voice.

"Hello," the man said.

"Hello . . . Mike?"

"Hello," he said again exactly the same way.

"I'd like to speak with Sally."

"She doesn't live here anymore."

"Oh," I said, feigning coolness. "Where'd she go?"

"Outta town. Got a job. She won't be back."

"How can I get in touch with her?"

An uncomfortable silence followed. This was mad Mike, annoyed at the stranger who had been sleeping upstairs in his house with his sister.

"Send her a letter," he said. "Takes three weeks if you hit the mail plane right."

"Mike," I said sincerely, almost pleading, "I don't mean to bother you, but where did Sally go?"

"McCarthy. We were fishing down there, and the day we left we stopped at the lodge for a good-bye drink. The owners offered her a job. Room and board. The whole works. "

"Is there a phone?"

"Radio phone. Restricted. Emergencies only."

"No phone?"

"Closest phone is Chitna, seventy miles away. Forget the phone! Write a letter." Mike gave me the address: McCarthy Lodge, McCarthy, c/o Glen Allen. I thanked him, but he'd already hung up.

I put down the phone, walked outside the shack, sat down on a picnic table, and began dealing with a terrible sinking feeling. Sally, vanished into the bush. Just as we were getting to the point when I finally had some time, she up and leaves the country. I'd be leaving soon for fires down south, gone six weeks—maybe more—and by that time who knows what might happen? At the worst, she'd find someone else to enjoy her summer with. At best? Maybe she wouldn't. In any case, I wouldn't be seeing her, talking with her, or making love with her for a long time.

"Fuck it," I mumbled. "I hate this job."

I spent the rest of the day trying to find out if there was any way I could get a message to Sally. The phone company did some checking and told me that there was a radio phone at McCarthy Lodge. It was restricted, they said. Emergencies only. Just like Mike had said.

I felt abandoned, lost. I had to talk to someone. But who? For sure, not Mitch. I'd made the mistake of seeking sympathy from jumpers before.

But that evening Charlie Brown came by my room. Charlie was different, and before I knew it I was telling him the whole story.

"Murr," he finally said, raising his hand to stop me. "Hold it, hold it. You need to get a grip. You know what I mean?"

Staring blankly at him, I wasn't sure that I did.

"Charlie, I know this sounds nuts, but Sally's pretty wonderful."

"That's fine," Charlie said. "But, it's no reason to flip out. If it's right, it'll still be there. If not, you had your good times." Charlie shrugged his shoulders as if to indicate the obvious utility of such a view. "Look at it as a gift. Accept it and be thankful for it."

"I'm going to miss her, don't you see? I was fond of the woman. I had my heart set on seeing her again."

"You can't set your heart on anything during fire season. You of all people should know that. You're the old-timer here, remember?"

"Thanks, Charlie, really, but I'm a little blue to wax philosophical right now."

"You'll be all right," Charlie smiled. "You're just bummed because you've been weaned off all that nasty sweet stuff. So, big deal. Now, you're in the same boat with the rest of us. Come on. Let's head out to the Howling Dog and have some fun."

No way was I going to the Howling Dog. I stayed in my room feeling awful and watched as Charlie drove away. The sky had clouded over and was getting dark. From the parade grounds of Fort Wainwright came the ten o'clock playing of taps. Across the empty courtyard, the mournful notes wavered in the wind, then echoed around the buildings to die in a reverberating silence.

A light rain began to fall. I lay awake most of the night tossing and turning, tormented by a nagging hopelessness that cycled between the thought that it was over with Sally and the painful recognition that when it came to love and women, I was a lost cause.

Out of the darkness boomed the voice of Rod Dow, self-proclaimed smokejumper authority on women and relationships.

"You're a bonehead with women," he said.

Rod insists that this is not merely an occasional affliction with smokejumpers but a rather universal one.

"And Taylor," he persisted emphatically, "you're one of the worst."

By morning it had come to me. I would go to McCarthy and find Sally. Chances were good that the first round of jumpers would be leaving for the Lower 48 within the week. I had to see her before we left or I'd be stuck the rest of the summer, tormented, everything still up in the air. Right after roll call I went to Tom Boatner's office and took a chair as he excused himself to top off his coffee cup. I sat there looking through the window and out across the aircraft ramp, inwardly thrilled by the winsome prospect of my new plan.

Boats came back in, closed the door quietly behind him, and plopped down in his swivel chair.

"What's up?"

I told him about Sally and that we'd probably be going south soon and that if I didn't see her now I might never see her again and that I really needed to see her and that it was a long way away and there was no phone and the mail only went in once every two weeks and sometimes not even that . . .

Boats grunted rather loudly, took a long sip of coffee, glanced sidelong at me, then back at his cup.

... and it would take a day to get down there and a day to get back and it was hardly worth going for less than two or three days and that would mean a total of five days and I knew it was still fire season but things had slowed down and it was raining ...

"All right. All right," Boats blurted out. "Call me when you get to Chitna. If anything's happened you'll have to turn around. If not," he said, smiling broadly, "have a great time, and we'll see you Monday morning."

In the back of my pickup, my waterproof pack rode under a tarp. On the seat beside me were two cardboard boxes. One of the boxes contained some of Sally's things from her brother's place, extra clothes, her art stuff. When I'd called Mike back and told him I was going to McCarthy to see Sally and would take some of her things to her, he got huffy. "Why don't you quit calling here?" he snarled. "You're just like a puppy over her." My interest in his sister clearly disgusted him. Reluctantly, he agreed to my offer.

In the second box I'd packed away a few items I picked up in town: a fluffy, peach-colored bath towel, two matching washclothes, some fragrant bath soaps, shampoo, film for her camera, an Enya cassette tape, two boxes of Kleenex, and two cartons of tampons. In McCarthy, such things are hard to come by.

The weather east of Chitna was warm and sunny with the coming of early afternoon. I rolled down my windows and the dust boiled in, settling on everything. It struck me as possibly ridiculous to be traipsing about in a vast wilderness searching for a woman I hardly knew, traveling four hundred miles to find I knew not what, beating my pickup to pieces, choking on dust, all the time thrilled that I'd soon be seeing her, yet fearful that maybe she'd already met someone else. It would be just like me to show up unexpectedly and become the laughingstock of McCarthy.

After twenty-six miles, I stopped at a wide spot along the Chitina River to stretch my legs. The water was swift, cold, blue-green, clear, and pleasant to the ear. Shaking the dust out of my shirt, I splashed up under my arms and dipped my head under. I toweled off with my T-shirt, then went to fetch another out of my pack. I sat for a while beside the river, enjoying the sun, part of a scene that suggested that I could be the last person on earth, perhaps even the first. In the sweep of my vision, Alaska offered a world devoid of any sign of human activity—only snowy peaks, gray stone ridges, and a river filled with sunlight racing before me. Beyond the far bank, as far as I could see, green and purple meadows ran to the foot of distant mountains, the canyons of which lay choked in the frozen blue remains of the last ice age.

At fifty miles, the road curved north and began following the Nizina River. Now and then through the trees there appeared what looked to be a small settlement about three miles east. The road stayed rough and slow but at last veered right toward the settlement and ran a straight line through a corridor of tall paper birch.

Emerging from the birch, I drove slowly into a large opening where several cars were parked on rough rocky ground. To the south, I noticed a few crude campsites and two dilapidated outhouses. Upstream a quarter of a mile, the Kennicott River erupted from a dark cavern that extended back under a mountain of ice—the Kennicott Glacier. Welling up and falling in treacherous cataracts, fiercely cold and laden with silt, the river sped along, freed at last from the glacier's icy darkness to rush into the light of a warm summer afternoon.

I sat listening to the river, mesmerized by the wonder of it, laughing inaudibly into its great roar, feeling happy, despite the fact that by now, word would have gotten around the shack. He's a mess, all right. The little redhead's put Old Leathersack on a fast track to the funny farm.

That's what happens when you get to be his age, Charlie would say. You know you'd better get yourself a woman while you can.

Hell no, others will argue, he's been that way all his life; never could tell love from a gopher hole. Saw him take off at night one time in a single-engine plane and fly over the Ochoco Mountains in pitch dark looking for La Grande because he'd met some half-Basque, half-Nez

Percé Indian princess. Shit, he got lost, almost ran out of gas, and wound up stuck in Walla Walla all night.

McCarthy was across the river two miles east. The bridge that once carried the railroad to the settlements of Kennecott and McCarthy had washed away, leaving the people who lived there and ran businesses no choice but to move their heavy items across the river in the winter when it froze. In summer, all goods and people had to cross using the aerial tram.

On each side of the river, a large deadman, a concrete block, had been buried in the ground to anchor a cable that spanned 275 feet of rapids. Log platforms had been constructed at each end to serve as both loading and landing docks. The little trolley had no doors or top, just an angle iron frame—sturdy but no frills—allowing barely enough room for two people and two backpacks. The trolley's two wheels traveled on a fixed cable, and it was propelled by pulling on a continuous circular nylon rope that looped through two pulleys, one on each platform.

Once across, I hadn't gone a quarter mile before I came to another river spanned by a tram system identical to the first—a different fork of the same river. The path led east, a narrow road bordered on both sides by dense willows and alders. The air was heavy with the smells of an Alaska midsummer afternoon: the bitter scent of cottonwood, the witch hazel smell of willows, and the subtle fragrance of wildflowers. In the distance, my ears picked up the patient chugging of a solitary diesel generator.

Up ahead stood a pale yellow building with a faded red roof. Once the railway station for McCarthy, it was surrounded by weeds, its windows broken and its boardwalk a shambles. Sets of rusted railway wheels lay strewn here and there between piles of steel rails. A sign on the side of the station nearest the road had one arrow pointing left to Kennecott, the other to the right for McCarthy. The Kennecott mines were named for the adjacent Kennicott Glacier, but a spelling error on one of the early claim maps was missed, and the spelling remains different to this day. Under the arrow pointing right, someone had scratched the words "McCarthy Lodge."

Past the station I stepped out onto a dirt street that reminded me of a scene from *Gunsmoke*. Weather-beaten buildings presented false fronts, hitching rails and wooden sidewalks; the most prominent building was surrounded by old wagons, wood-spoked wheels, and steel ore cars. Exceptionally tall aspens grew in a line along the west side of what must have once been Main Street. At seven in the evening, they cast long shadows across the street and onto weathered gray-brown walls. I felt my heart racing and a sudden rush of doubt. I had chosen to up the stakes, and now it was time to let the cards fall. An old black dog came waddling off the wooden sidewalk, smiling and wagging its tail.

Ahead to my right, I saw a sign:

MCCARTHY LODGE
WELCOME TO MCCARTHY, ALASKA
GOOD FOOD   COLD BEER   HOT BATHS   ROOMS

Stepping inside the entryway, I was greeted by an elderly man and woman sitting at a round oak table. The man was wearing a T-shirt that said WHERE THE HELL IS CHITNA?

"Hello. Can we help you?" the woman asked.

"Well," I said. "I hope so . . . I'm looking for Sally."

"Sally?" she said. "She was just in the kitchen a minute ago."

Before she'd finished, Sally came backing through the twin doors holding a tray of white cups and saucers. She set them on the nearest table, then looked up.

"Murry," she squealed. "Murry! I can't believe it."

Sally came running, flourishing a big smile and throwing her arms wide.

"I can't believe it's you!"

We hurried outside, where Sally gave me a big hug and kiss and threw herself back at arm's length.

"You got my letter? No! You couldn't have. The mail plane hasn't even come yet . . . How did you know?"

"Mike told me. We'll be leaving for the Northwest soon. I had to come."

Sally gave me another hug, then had to get back to work. Back inside she introduced me to the two people at the table. They owned the place. Dinner was soon to be served, and the guests at the hotel would be piling in. I told her that I had to go back to the pickup and get the things I'd brought for her.

"I'll serve you dinner when you get back," she said.

Soon I was on my way down the road through the woods to the river crossings and on to my pickup, humming happily. I got the rest of my gear—a tent, a sleeping bag, some food—and the gifts for Sally, and stuffed them in my pack. Having driven four hundred miles, including the seventy from Chitna, I'd walked six miles and made a total of six river crossings by the time I got back to McCarthy.

After a shower and shave in the hotel, I put on clean Levi's and a fresh shirt, stashed my pack at the end of the hall, and took up a table by a window that looked out onto the street. From that vantage point I could see the mountains, the town, and Sally as she went about her duties.

The half-dozen tables were occupied by mountaineers from Germany, Switzerland, and Austria, young men and women of adventurous spirit, sunburned, woolly haired, and shod in rugged hiking boots. Sally seemed in her element serving them, giggling and exchanging reassurances about language barriers. She wore the Lodge's required attire, a suspendered floor-length navy blue dress and a white blouse with dainty ruffles at the sleeves and neck. Pinned to her hair was a bonnet into which she had tucked a few daisies. For the moment, she was perfectly Amish. Swirling from table to table, Sally tossed me an occasional smile. There was no menu. Whatever they had for that night would be served family style. I decided to sit back, wait awhile, and take in the scene.

While the outside of the lodge was not stately, the inside, in a rustic fashion, was. The main dining area had heavy round oak tables, antique chairs, and was floored with heavy planks, oiled and foot worn. Tall windows and doors ran to a comely height in relation to the vaulted ceiling. Along the high walls hung antique mirrors, clouded by time and dust.

What I saw next at the back of the room was as unexpected as it was lovely—a grand piano. I went over, sat down, and began playing. Sally came out of the kitchen with a frosted mug of beer and handed it to me.

"I didn't know you played," she said.

After an improvisational medley and some encouraging looks from the guests, I played John Lennon's "Imagine." Sally swirled back and forth between tables with her food trays held high as the music provided a background to her new life.

By eleven P.M. Sally had finished cleaning up in the kitchen. She went to her room and gathered her things for our hike and met me behind the hotel.

"We can't stay here, you know. It's their home . . . it wouldn't be right. Are you ready for some walking?"

I was and we did—all five miles up the old road to Kennecott. I carried most of the gear in my jumper pack-out bag. It was uphill all the way but not too steep. We talked and stopped here and there for a breather and a kiss. To the west the evening sky flushed pastel yellow and orange, lit by a sun traveling sideways behind sixteen-thousand-foot Mount Blackburn, a striking backdrop for the silhouettes of birch and aspen.

At that hour the birds were quiet, the air lay still, the only sound that of distant rushing water.

For me it had been a long day, and I was ready for it to end. But what I saw next startled me even further. Looming before us, like the hulk of an enormous ship, were the crumbling ruins of the processing plant for the Kennecott copper mine. On a slope running uphill to the right, the structure towered eight stories on the left, two at the upper end. This weather-beaten monster stood over a hundred feet tall and was three hundred feet long. It hovered in the soft morning light like some defiant creature that had crawled up out of the bowels of the earth, gotten into a terrible fight, and been left to die a slow death. Boarded sides had been painted dark red, the window frames white. The windward upper corner had been ripped away by ice storms, leaving a gaping hole exposing the skeletal remains of massive Douglas-fir beams, collapsing floors and stairways. Cables, pulleys, electrical wires, and trolleys hung in tangles like the entrails of a great robot. To the right, at the east end, a forty-foot-diameter horizontal drive wheel stood motionless, its cables stretching three miles up the mountain to the top of Bonanza Ridge and the Mother Lode, Jumbo, Erie, and Glacier mines.

Beneath the crest of Bonanza Ridge, near a second wheel, were several similar but smaller buildings. These upper ruins took their footing on steep hillsides and stood five stories tall, pressed against vertical cliffs. Abandoned for over fifty years, no sign of activity remained, just weathered structures taking on the character of the lichen-covered rocks to which they cleaved—brown, burnt-orange, yellow, and gray.

"My god," I said, turning to look at Sally.

"Isn't it wild? To think that so much happened here, and now it's all gone. All those people and their dreams and now it's just wreckage."

Large cast-iron drive wheels up to sixteen feet in diameter stood before us, resplendent and worthless. Attendant to the main processing plant was the rest of what had made up the town during its heyday. We passed the dairy barns and the meadows where the cows had grazed on summer grass. In winter they fed on imported alfalfa hauled in from Washington State on the train.

"Kennecott is where all the work was done," Sally said. "McCarthy is where everyone went to play."

We followed the railroad tracks across a massive wooden trestle until they stopped a hundred yards beyond the archway formed by an overhead bridge.

"I know a good place," Sally said, shading her eyes from the sun.

A quarter of a mile from the plant, an old wagon road became a trail. A quarter mile up the trail we stopped. Root Glacier was two hundred yards away. To the east and north was the broad expansive slope of Bonanza Ridge, green along the bottom, light brown at mid-slope, and gray-white with rock and snow toward the summit. The sun flashed on a few remaining windows of the upper ruins. How strange they looked, the falling-down buildings, such fantastic structures wedged into a fortress of stone high in the sky.

"This is it," Sally said, taking a side trail up a small bluff. Soon we reached the brim of a small flat where a dozen scrubby spruce formed a circular enclosure. Down in a cozy hollow was some flat ground covered with spruce duff. It was just enough space for our tent, a woodpile, and a campfire. The entire area could be covered by my ten-by-ten-foot blue rain fly.

We unloaded our packs, stuffed them under the edge of the tarp, and crawled into the tent. At last Sally and I lay side by side in the rosy glow of the tent's interior, naked on our sleeping bags, deep in the wilderness. Not far away the glacier shoved and groaned, sending shivers through the earth.

In that moment, all was present and immediate, everything whole and complete. I was with Sally. I held her in my arms as she held me in hers and we kissed and giggled and made love until, at last, our bodies trembled like the earth beneath us.

We slept until eleven that next morning, then went for a brunch of tea and homemade pastries at the newly renovated Kennecott Lodge. Lounging in the sun on the veranda, we talked softly and looked out over the big valley on a bright and cloudless day. Most of the afternoon, we spent exploring Root Glacier. Afterward, drowsy from the warm day, we went back to our camp and took a nap, laughed and petted at length, made love, fell asleep, woke up, and started pawing at each other all over again.

"I'm glad you came," Sally said softly. "I wanted to share this with you. Everything I saw here made me think of you and how much you would like it."

Sally raised up on an elbow and turned to me, cradling her chin in her hand, her eyes smiling, and her freckles so beautiful they were almost indecent. Ringlets of damp red curls fell forward across her face.

She looked at me a moment, then kissed my nose.

"You're such a sappy romantic, Mr. Taylor. But, I have to admit, I like you."

"The word incurable comes to mind," I told her. "I do tend to get carried away."

"Well," Sally said, before giving me a long kiss, "you can get carried away with me anytime you want to."

For the remaining two days, Sally and I spent the nights and mornings together, then she went to work and left me to work on the pickup and explore. About seven each evening, I'd make my way to the lodge, take a hot bath, put on clean clothes, chat with guests, play piano, drink cold beer, and wait for Sally.

One evening we walked up to where the river came rushing out from beneath the Kennicott Glacier. Sally sat down on a gravel berm, elbows resting on her knees, the river wind tugging at her red curls.

My time with her had been so short. We'd had little time to do much more than meet briefly, give in to the need for intimacy, and then, in my case, yearn for her constantly. What could she possibly mean to me? What I felt was crazy, but as she sat there, her exquisite profile lit by a fading sun, I knew that I was falling in love with her.

Our last evening in McCarthy, Sally and I walked to the edge of town along a set of tire tracks that skirted the rim of a small bluff. We lay in a grove of aspens, milk-green trunks under clouds of rustling leaves, and started kissing and petting. Then the mosquitoes roared in, and we had to get up and move on.

The road went another quarter mile, then intersected the south end of the McCarthy airstrip.

Just before we arrived on the edge of town, I led Sally along a game trail that ran east about a hundred yards. I'd chopped away the lower branches of an old white spruce to make room for our tent underneath. The entrance was close to the trunk, and the scent of fresh spruce sap saturated the air, tangy and sweet.

That night Sally's desire to touch and be touched was so wanton and unrestrained that I think she may have felt a little embarrassed by it.

"Who am I to deserve this?" I said, my lips pressed to her upside-down tummy.

"You deserve everything. More than everything."

She switched around and kissed me fully on the mouth like she couldn't get enough. She straddled me, put me in her, chatted softly to me, explaining things. She excited herself with the talking. Her agitation became almost violent, close to hysteria, not far from pain. Then she seized on me, crested her love wave, and let out a mournful sigh. Finishing, she got the giggles, collapsed on top of me and burst into tears.

We lay still for a while, our bodies relaxing, Sally's breasts against my chest, our legs entangled, hearts as close as two can get.

"Meeting you has been the best thing of my whole summer," I said. "Leaving you is going to be hard. I'd love to stay, but I can't."

"I know," she whispered. "You have to go off to your silly old fires."

I lay awake a long time holding a sleeping Sally in my arms, feeling her close, feeling her warm breath. I knew I was living a moment I would always remember. I tried to visualize the surrounding mountains and sky and forest as they would appear from jump altitude. From that distant and somewhat detached perspective, I could see that at times my life was like a fine piece of art. But I also knew that life, unlike art, is not perfect. With all its power, dreams, and dear hearts, for every precious moment, life remains tenuous and uncertain.

16

The last bunch of the down-south boosters were airborne by 5:30 in the morning. Some of their faces must have looked like they belonged in a morgue. Not just because they had to be up at 4:30 for their ride to Fairbanks International but because they'd been out all night, drinking and carrying on. Homeward bound, they flew high above the clouds, slumped in their seats, reeking of booze and bar smoke.

It all started around quitting time the previous day. The tension had been building all afternoon. Word had come down from the main dispatch office, the Puzzle Palace, that the remaining boosters were being sent home, so the date for the Big Flip was moved up to include them.

Just before six o'clock an announcement was made over the PA system.

"OK! We're going to start the Big Flip about seven," said Chip Houde, manager of the Alaska Smokejumper Welfare Fund. "Out on the ramp if it's sunny, in the paracargo bay if it's not."

A hearty cheer erupted.

"We'll go over the rules just before we begin," Chip said, raising his voice to more cheering. "In the meantime, free beer, free burgers, and free chips . . . courtesy of your welfare fund."

Another great cheer went up, and the crew headed for the ramp and some serious socializing. Paracargo hand-dollies rushed by, loaded with cases of Meister Brau and Old Milwaukee. Two large barbecues started belching black foul-smelling smoke across the airfield. Cardboard boxes filled with frozen hamburger patties and hot dogs appeared on picnic tables. Dogs and burgers were quickly tossed onto the grills. Bags of potato chips as big as fifty-five-gallon drums were ripped open and pounced upon. Several burgers burst into flames and had to be doused with beer, but the fire was too hot. Reigniting, they hissed and shot flames two feet tall as they began melting through the grate. Others were snapped up and placed between semi-frozen buns.

"Get the hose!" Erik the Blak croaked between hearty gulps of beer.

A general complaint began to circulate about the smell of the smoke.

"Smells like creosote," Mitch said. Heads turned and burger eaters froze, jaws suspended on the downstroke, eyes fixed on a pile of railroad ties the yard crew had been using for landscaping.

"Ahhh, creosote!" Kubi wailed. "Who put that shit in the fire?"

Two rookies came running up with shovels and began digging the larger creosote pieces out of the barbecues and carrying them out to the gravel driveway. A few smaller chunks had fallen from the barbecues and lay on the tarmac, puffing away, before being stomped out in a flurry of contempt.

"Who the hell put railroad ties in the barbecue?" Seiler demanded.

"Probably rookies," Mitch said, scowling. "Who else is that dumb?"

The air filled with laughter, loud talk, and the sound of empty beer cans clanking into garbage cans. Burgers, hot dogs, and steaks were salvaged from the flames and flopped onto paper plates, while discarded potato chip bags lay here and there, trampled underfoot. A few pretty girls showed up. Erik the Blak's voice had gone up ten decibels—the party now fully under way. A half hour later the Big Flip began.

"OK! Listen up everybody," Chip Houde yelled, thrusting his arms into a V over his head. "It's showtime. Everybody line up at the paracargo bay. Let's get this thing *going*."

"Dow for seven, Dow for seven," chanted the group as Rod Dow stepped before them, his hands signaling for silence. Once again the sacred origin of the Big Flip was to be recounted.

After more than twenty seasons, by a rather bizarre and barely credible set of circumstances, Rodney Irving Dow has managed to become the least-promoted smokejumper in history. Back in the early eighties an underground campaign was spawned within the crew to somehow get Rod promoted from the entry-level GS-6 he'd received in 1968, up one grade to a GS-7.

Smokejumpers everywhere took up the cause, writing DOW 4—7 around the world. A photograph hangs in the ready room of the standby shack showing Boats and Buck Nelson atop Mount McKinley holding up a small banner with "DOW 4—7" stenciled thereon. In China, a jumper photographed the words dabbed in black paint on the Great Wall. People flying over Peru have seen it written on the plains of Nazca. On the rest room walls of the world, only one other sentiment has been expressed more frequently. From Burns Brothers truck stops in eastern Oregon to the white sands of Bora Bora, this elegantly simple appeal can be found petitioning the universe for Rod's advancement.

Despite such effort, Dow remains a GS-6. Twenty-three years without a promotion would discourage most people. Not Rodney I. Dow. Why should it? Never once has the man applied for a promotion of any sort. When management has encouraged him to take one, he just possum grins at them and declines.

Despite the bleak history of his career, Rod remains the undisputed authority on such esoteric matters as Big Ernie and the Big Flip.

"It all started back in '68 or '69," Dow said, referring to his first years jumping out of McCall, Idaho. "Hell, I can't remember. That was a long time ago. Anyway, we were building some corrals up at Lake Creek.

"It was in the afternoon. There hadn't been many fires, and we were bored with the project—*really* bored. The day was completely overcast. Suddenly the sky opened up, and the words BIG FLIP appeared, spelled out in little clouds. I know this is hard to believe, but it's the truth. It was totally amazing."

The group cheered their approval of the story, true or otherwise.

"But," Rod went on, "we couldn't understand what it meant. Maybe, we thought, it meant that we should take one of the guys and flip him in the air. But how would we know which one to flip? Someone said, 'Let's flip a coin to see who it'll be,' and that's when we got the message. It sounds unbelievable, I know, but that's exactly what happened."

In the years that followed, the McCall jumpers traditionally had their Big Flip near the end of fire season when the crew was flush with money. When Dow transferred north to Alaska in 1974, he brought the idea with him and spent weeks promoting and organizing the first Alaska Big Flip. Alaska jumpers, typically a tight-fisted lot, were nearly hostile in their skepticism of the newcomer from Idaho.

"Oh," Rod would assure a prospective flipper, "I think your chances of winning are especially good."

The first Alaska Big Flip was conducted in accordance with official McCall rules. Just the same, it left a sour taste in everyone's mouth—because, after convincing a total of thirty-two jumpers to invest twenty dollars apiece, Dow walked away with all the money. Infuriated, the losers wanted Dow's hide, even though they'd witnessed with their own eyes his fair-and-square flipping of seven straight tails in seven consecutive rounds. As it turned out, Dow gave a third of his winnings to the welfare fund and threw a rip-roaring party with the rest. The Big Flip has been an annual event ever since. Over the years, the stakes have been steadily raised. The largest sum to date being the sixteen-hundred dollars won by Scott Dewitz, an Alaska jumper on detail to McCall in 1989. This year, with the big season and lots of overtime, the entry fee was set at fifty dollars—another of Dow's ideas. The flip we were about to witness would be the richest in Big Flip history.

One by one the jumpers lined up outside the paracargo bay, wallets in hand. Houde and Dow stood on each side of the door, Dow on the left taking money, Houde on the right writing down names.

With the last of the line inside, the count stood at eighty. The door to the paracargo bay was quickly locked, the windows taped so no one could see. Dow spread the money out evenly on a table near the flipping arena, fanning the twenties, fifties, and hundred-dollar bills like playing cards into a line two feet long.

"Eighty times fifty," Rod called out. "*Four thousand dollars.*"

A roar flew up in the paracargo bay as the crew scrambled for seats. Some sat on paracargo bins, others atop rows of fire packs. The rest took their places on pallets of containerized chain saws, fire pumps, and hose. The flipping arena consisted of a platform the size of a pool table. A six-inch wooden frame elevated the edges, over which a piece of bright yellow cloth had been stretched tightly. Each player was to stand at a blue line painted on the floor four feet from the end of the table and was required to toss the official quarter through a small upright shaped like a goal post. After landing on the cloth the coin would be observed by the judge and announced either 'Tails, you're in' or 'Heads, you're out.'

Houde raised his hands for silence. Just then the door to the shack flew open, and a cargo dolly with two more garbage cans of iced beer was hastily wheeled in. A great cheer went up.

"There aren't many rules to the Big Flip," Houde yelled. Another great cheer. "So listen up—we don't want any whining about not understanding the rules." More cheering.

"Rule number one. No stepping over the blue line. If you do, *you're out!*" Chip was emphatic about the "you're out" part, yelling it with the finality of a home plate umpire.

"Rule number two. If the quarter doesn't go over the goal post and it's a head—you're out! If it's a tail—you get to throw over.

"Rule number three. If the quarter hits the table, bounces or rolls off, and it's a head, you're out. If it's a tail, you get to throw over."

Every rule was met with a hearty, roaring cheer. Eighty grown men behaving like kids.

"Rule number four. During the first round you may buy back in for fifty bucks, as many times as you like. But once the first round is completed—no more buy-backs!" Chip froze for an instant, eyes bulging. "And that's the rules," he yelled.

Again, a great roar.

Chip lifted his hands, this time more ominously.

"Let the games begin. The first players are . . . Woods, Persons, Seiler, Troop, Robinson."

When my turn came I was a bundle of nerves, so desperately did I want to win all the money in that lovely green pile. I promised myself that if I won, I'd charter a bush plane and fly to see Sally. Picking the coin up off the table, I held it to my heart, looked at the ceiling, closed my eyes, and silently pled my case to Big Ernie that if he had any hair on his ass, he'd see this as his big chance to reward me for all my hard work and unfailing loyalty.

I flipped. Chip lurched forward, eyeing the coin.

"Good flip," he cried. "Heads—*you're out!*"

I staggered back, struck by the horror that I should have known better than to try to kiss up to Big Ernie.

Chip's "you're outs" were harsh, final, and instantly incited a great cheer. But when someone flipped a tail, Chip's demeanor was that of a man witnessing a miracle. Heads or tails, the crew roared with equal enthusiasm.

The first round finished with about half going out. When it came Chip's turn, everyone braced themselves. He closed his eyes, whispered something to the ceiling, and tossed.

Dow jumped forward, eyeing the coin eagerly. "Excellent throw! Heads—*you're out!*"

Chip staggered back with the news, but by the time the cheering had subsided, his wallet was out and he was brandishing a new fifty-dollar bill.

"I'll buy back," he cried.

The crew cheered their approval, yelling, drumming loudly on fire packs as if berserk. Chip repeated his ritual with the ceiling. Once more he tossed the coin. Dow eyed it sharply. "Excellent throw! Heads—*you're out!*"

Chip went shark-eyed, dug out another fifty dollars and tried again. Same results. Wobbling in defeat, he babbled something to himself, and returned to his position as judge.

Scott Dewitz, indisputably smokejumping's most despised flipper, tried a buy-back as well. Dewitz had already won two Big Flips, and the crew loathed the possibility that it could happen again. Upon winning sixteen hundred dollars in the Big Flip at McCall in 1988, Scotty had stood on a chair, waving a fistful of bills over his head, gloating, and yelled loudly, "I . . . am a winner, and you . . . are all losers." Now, as he stepped to the line, the crew unleashed a roar of booing and insults. Scotty smiled, unaffected, closed his eyes, held the quarter to his lips, blew on it, and tossed.

"Good throw!" Chip roared. "Heads—*you're out!*" The luckiest flipper of all time dragged himself off, hunchbacked, cringing under a din of cheers.

That ended the first round. No one wanted anything more to do with buy-backs.

The second round eliminated about half the flippers again, leaving about twenty players. The third round left thirteen, the fourth only five. The paracargo bay vibrated with suspense. The fifth round eliminated no one; all flipped heads. One tail would have won the entire pot. The sixth round left two players, and the cheering stopped.

In the seventh round both players flipped heads. In the eighth, both flipped tails. In the ninth, Geoff Curtis, pilot of *Jumpship 19,* flipped a head. Kind, mild-mannered Dave Estey, a Boise jumper, followed with a tail.

"Yahoo!" Dave yelled, a split second before he was drenched with a barrel of ice water. Totally soaked, his face clammy with shock, Dave gasped for air as Chip presented him with a bundle of bills three inches thick.

"The final sum," Chip announced, "including buy-backs, comes to four thousand four hundred dollars."

Dave waved the money happily over his head. When the crew finally calmed down, he had something to say.

"First I'll take back my fifty dollars. From the rest, one-third will go to the party tonight at Pike's."

Not surprisingly, this action met with universal approval and brought on another blast of barbaric applause. The gesture of donating one-third to a party was traditional, but what came next was not.

"In addition, I'd like to share this with others less fortunate than ourselves. I'll split the rest—on behalf of the BLM Smokejumpers—between Billy Martin's family and the Fairbanks shelter for the homeless."

The cheer of all cheers followed. Howls, hoots, chimpanzee wahbarks, whistling, yelling, and drumming rocked the paracargo bay, as the crew barreled out the door and headed for Pike's Landing.

As our mob filed into Pike's a half hour later, Chip was busy negotiating two thousand dollars on account at the bar. This was more than one-third of the prize money. Apparently, Estey had been talked into a decrease for the homeless and an increase for the hapless.

The party escalated full tilt. The jumpers began to enjoy themselves in a lost and savage way, the knots of responsibility loosened for the night. Two changes in conduct occurred almost immediately. For one, our taste in alcohol suddenly went from Meister Brau and Old Milwaukee to Crown Royal and Cuervo Gold. Some of us stayed with beer, but it was Heineken, Alaska Amber, Moosehead, and Dos Equis.

Jumpers gathered on the deck, laughing, joking, and listening to Don Bell tell rollagon stories.

Over the years Bell had developed what he called a "a strange and rather violent relationship" with rollagons. Rollagons are rubber fuel bladders shaped like fat tires. Two sizes are used by the fire service— 250 and 500 gallons—four and five feet in width, respectively, and three feet in height.

Rollagons do two things particularly well—weigh a lot and roll. Heavy enough to mash small objects, they eventually react to gravity and begin to roll. They roll so well, in fact, that they can move great distances on what appear to be perfectly flat surfaces. Left unattended, on the slightest incline, rollagons take on a life of their own and soon are on their way. It was during the years when paracargo made a lot of aerial fuel deliveries that Don and the otherwise righteous-looking, black center-hubbed rollagons became mortal enemies.

"Rollagons are evil," Don had told us many times. "They're possessed. They were put here on earth to do harm. I keep an eye on them. I keep them in their place."

Bell would fiercely disagree, but everyone will pretty much tell you that he has had marginal success at keeping them anywhere.

The rollagons' first great escape came when a three-thousand pounder filled with Jet A rolled off across the airfield at Bettles, disappearing into the sunset like a landlocked whale. By the time Bell spotted it and caught up with it, it was approaching the edge of a steep embankment. Don tried to chock it with rocks, but the rollagon pressed hard against them, mashed them into the gravel, and then continued on its way. As the rollagon reached the edge of the embankment, Bell grew frantic, cussed at it, and kicked gravel at it. Finally, he wound up and gave it a rabbit punch that sent him howling in circles with two broken fingers as the rollagon plunged over the bank and tore into a thicket of willows.

Defeat came again when Don was sent to Lake Minchumina to set up a fueling site. As he inched the "rollie" down an inclined ramp, it rolled onto his foot and trapped it. Don pushed back desperately, but as his strength gave out, the rollie levered Don to his knees, forced him to the ground, then bowled over him, barely missing his head, and left him on all fours, groaning and cussing. There was no one else around, so Don had to crawl into the brush and spent the night with a throbbing knee, swollen tight inside his Levi's. The next morning, mosquito-bitten and weary, Don dragged himself down the airstrip to the BLM fire cache, got on the phone, and called Fairbanks for help. Back in town, Don was hospitalized for a week with various abrasions, damaged ligaments, and torn cartilage.

"I've had a fair bit of trouble with rollagons," Don admitted to the group on the deck at Pike's. "But the worst thing that ever happened was the roll-a-bomb!"

He immediately let out a hyena yowl, gyrating around, hands flying. "The roll-a-bomb," he cried, barely able to contain such delight. "We were making a fuel delivery to this big fire in the C-119. We had several rollagons on board, mounted on plywood pallets and all lined up on the roller track. When we got to the fire I called Al Dunton, the fire boss, on the radio. He told me that they'd been chasing the fire all week and that they had finally stopped it at the river, and if they could only hold it for another day, they would have it contained at thirty thousand acres. Dunton was the head of the jumpers at the time, so naturally we wanted to impress him.

"Al told me the best place for the fuel was right across the river in a big meadow. I remember watching the first rollagon as it tumbled off the end of the roller track. The static lines pulled free and trailed in the wind under the tail. Both parachutes were sixty-four footers. The first one came out in a big messy ball. The second one strung out straight as a stick and stayed that way."

Don's eyes grew big, his hands poised before him as if he were looking into a crystal ball. Whistling the sound of falling bombs, he suddenly flung his hands wide and made a terrific exploding noise.

"I saw it hit. A big silvery splash just before the release misfired. I never saw such a thing. This huge fireball, the size of a football field, blew into the air. The pilot looked back and screamed over the radio. 'What the hell is that?' 'That's the rollagon,' I told him. On one side of the river the fire was barely smoking and on the other side was this enormous fire taking off out across a meadow and right into a thick stand of black spruce."

"Boy, back in fire camp all hell broke loose. The overhead team started running around mad as hell. They had to fly people by helicopter and get them across the river. The fire was two thousand acres before they got it stopped."

Two hours later the open tab had been consumed, and our interest in Pike's Landing waned quickly. We stampeded out the front door into the parking lot. Engines roared, headlights flicked on, rear wheels spun. Pulling out of the parking lot, I saw Scotty Dewitz on the front lawn, down on all fours, puking.

I turned right off Airport and found myself heading down Cushman with a load of jumpers in the back. In my rearview mirror all I could see was a bunch of hairy-headed muppets silhouetted in the following headlights.

South Cushman Avenue holds a bizarre collection of business ventures. Starting at Airport Road, there's the Greyhound, the Drop Inn Cafe, Rocket Surplus, Discount Liquors. Farther south is an old-time barbershop, Alaska Sportsman's Mall, a couple all-night gas marts, three or four biker bars, and Cheap Charlie's. It's the kind of place where you can get a hair cut, get drunk, dance, grab a bite to eat, pitch your new camp tent in the bushes, and sleep the whole thing off. From the looks of the people you run into on South Cushman, many of them have apparently spent a good part of their lives doing that very thing.

Up the street a small marquee stood out in front of a large warehouselike building. One word, plastic black on plastic white—REFLECTIONS.

Once in the parking lot, my load of miscreants departed for the bar with such haste they left the back of the camper shell open and the tailgate down. Beer cans had rolled out the back and onto the ground. At that point, for whatever it was worth, I decided that I had best volunteer myself as a designated driver.

An unmarked door led into a small plywood foyer, painted dull black and lit by a bare lightbulb suspended from the ceiling. A hallway, also black and bare-bulbed, routed me to a window, where I paid the two-dollar cover charge. Through the next door, drabness blossomed into a realm of color, music, and swirling lights.

"It's all a matter of mirrors," Chip Houde once said of Reflections. Both the north and south walls are covered with them, the ceiling also, the wall behind the bar as well.

It wasn't hard to locate my charge of red-eyed whelps—they had taken up a number of the barstools within arm's length of the stage.

The music pounded loudly, making conversation impossible. My companions cast reassuring glances at each other, acknowledging that they had come to the right place. A line of tiny blinking lights ran around the rim of the stage, along walls, around the bar, and up and over the doors in a dizzying array.

Black light flooded down from the ceiling, turning the whites of eyes and T-shirts a glowing purple, human skin a rich dark brown. Purple foam topped fifteen-dollar pitchers of beer. Between long gulps, the spirited jumpers licked their lips and swiveled their eyes around center stage. The music started, hard, driving, and loud. A chorus blared forth . . . "We Don't Need Another Hero." Up a short stairway at the far end of the stage, four naked young women made their advance. A grating, hard voice rasped out over the PA system. I looked around through heavy cigarette smoke and saw a fat, bespectacled, bald man with a set of enormous gray whiskers, perched on an elevated platform filled with sound equipment, punching buttons with one hand and holding a microphone in the other. Eyes crossed with intensity, he conducted an experiment in special effects.

"O—K . . . my . . . fr—iends," he chanted, his voice rising and falling on each syllable, "Let's . . . give . . . these . . . la—dies . . . a . . . big . . . hand!"

The women danced into place along each side of the elevated stage.

"Let's . . . not . . . be . . . rude. These . . . la—dies . . . are . . . here . . . for . . . you. On . . . stage . . . left . . . wel—come . . . Glo—ri—ous . . . Glo—ri—a." Applause, yelling, mayhem. "On . . . stage . . . right . . . beau—ti—ful . . . boun—ti—ful . . . Bam—bi."

With each introduction a dancer would perform some sensual moves on a brass pole at center stage. Once the introductions were completed, the music was turned back up, and the dancing began.

The four women onstage began to writhe, lowering their bodies on bent legs, kneading their breasts, and looking my obstreperous pups in the eye. Wallets materialized. The dancers lowered themselves down onto the stage, rolling their pelvic regions. Breasts, butts, and pie-shaped patches of hair moved in and out and around. At differing angles, where mirrors mirrored mirrors, these bewitching females and their body parts appeared in a string of reflections that stretched to the edge of the universe. The jumpers crowded the edge of the stage like salmon waiting a waterfall.

At three o'clock I lobbied for leaving. We hit the door near penniless. The sun was high in the east as we made for my pickup. When I opened the door, I found what appeared to be a dead man lying across the seat. He jolted awake—a Lower-48 jumper. After levering himself upright, jelly-eyed, he sat staring out the windshield, swaying back and forth, and thanking me repeatedly for providing him a ride to the barracks.

The morning after was a gruesome affair—especially for those departing. A lot of them hadn't gotten home until after 3:30, and they had to be up and over to the shack by 4:30. Eric Pyne had crashed all night in the lounge. It was his job to make sure they didn't miss their ride to the airport.

"It was pretty bad," he told us later. "Two of them had to be rolled out to the bus on paracargo carts."

Pyne, gray-lipped and pale, sported a major case of red-eye himself. His hair stuck out in all directions as if it were trying to escape from his head. Eric conducted the morning roll call quietly.

"Dewitz . . . Anybody seen Dewitz?"

"The last I saw him," Romanello said, "he was fertilizing the lawn down at Pike's."

"Robinson . . . Anybody seen Tyler?"

"Yeah, I saw him about four o'clock," Chip groaned. "He was down at Reflections buying table dances from Cha Cha Rodriguez."

"Dow. Where's Dow?"

No one had the vaguest idea about Dow. It hadn't been a good night for brain cells. Half the crew was missing. The other half grumbled back at the box when their names were called, then fled in search of places to hide. Regular PT and work were not a priority for the day. We were free to sleep, rest, watch the tube, or pursue other personal detox programs. The day was windless, gray, cool. Ten minutes before quitting time, the siren went off and the first load—I was first man, third stick—rolled on a reported fire just south of Lake Minchumina. The call turned out to be a false alarm, so we patrolled north to Tanana, east to Fort Yukon, then back in to Fairbanks and were off at eleven o'clock.

17

July 29   Fairbanks

The last four days have been cool, cloudy, and quiet. Initial attack
has come to a standstill. The atmosphere around the shack is one of tedious resignation. Dispatch held three loads on standby for the weekend—a surprise to everyone. Yesterday at 6:00 P.M., two loads rolled to
a fire just north of Venetie. Only one jumped.

Talk is going around about the first down-south detail, with possibly fifteen to McCall, Idaho, or Missoula, Montana, and perhaps another five to Redmond, Oregon. No one really knows anything for
sure. The rookies and snookies are convinced that fire season is over.
It's a quiet Sunday. People write letters and phone families. Wives and
girlfriends drop by. A few jumpers have perched themselves in the
lounge, gawking at the TV like vultures eyeing a roadkill.

For the end of July the number of reported injuries seems about
normal considering the amount of activity we've had.

1. Rick Robbins: Fractured vertebrae, lower back. Still on light
duty. Prognosis: Will never jump again. (Early in the year, during a
rookie training jump, Rick brushed a tree, tumbled out, and broke
his back. He made an admirable attempt to hang on, but in the end

the doctors had warned of the risks of paralysis and suggested he give up his lifelong dream of becoming a smokejumper. Rick had to sit on the sidelines during a busy season—something any jumper would hate—and then, in the end, after all he'd gone through, he had to accept that he wouldn't be coming back. Rick's last days at work were sad, both for him and for us.)

2. Charlie Brown: Chronic lower-back pain. Went for X rays today.

3. Jeff Bass, Boise: Chain saw cut in calf. Cut to the bone. No permanent damage.

4. McCall jumper: Chain saw cut in calf. No bone damage.

5. Manny Diaz, Boise: Wrenched knee; stretched ligaments.

6. Damon Jacobs, Boise: Hyperextended ligaments in knee. Off the jump list for a while.

7. Al Seiler: Concussion; hit a tree on landing. Splinter in eye. Off the list.

8. Chris Cochran, Redding: Injured back. Returned to Redding.

9. Ray Brown, Boise: Twisted ankle running to catch spot fire.

10. Tom Boatner: Gimped knee, torn meniscus, arthroscopic surgery.

11. Bob Hurley, another Boise jumper: Injured in the Yolla Bolla Wilderness in Northern California. (Bob had difficulty slowing his canopy down, came in fast, and slammed into a downed snag. A branch speared through his jump jacket, puncturing his chest in the upper left quadrant. He was flown by helicopter to the trauma center in Redding, where X rays showed the lung to be OK.)

A study undertaken in 1986 examined the frequency and character of smokejumper injuries nationwide. During a seven-year sample period, careful records were kept of all jumps and associated injuries. In addition, an orthopedic surgeon and his staff performed musculoskeletal examinations and strength tests on twenty-eight randomly selected Alaska smokejumpers entering the 1987 season.

The 31,875 jumps of the seven-year period resulted in 267 injuries, an incidence of 0.8 percent. These injuries were sufficiently severe to require medical evacuation in 39 percent of the cases. In 31 percent of the cases, work capacity was temporarily limited.

From the seven-year comparative analysis and the 1987 musculoskeletal examinations, the general condition of the average smokejumper was summarized as follows:

75 percent had "worn" kneecaps.
53 percent had injured knees.
50 percent had injured their ankles.
43 percent had injured their backs.
28 percent had injured their shoulders.
26 percent had been knocked unconscious at least once.
25 percent had hamstring tightness.
11 percent had torn knee ligaments or menisci.
11 percent had patellar tendinitis.

In the fall of 1987, when the team from the sports-medicine clinic came to discuss the results of the study, we gathered in the lounge. They brought along videos and charts, and Dr. Carey Keller spoke.

"This has been an interesting study from the standpoint of injury therapy," he said. "Frankly, I've always liked you guys. I enjoy working with you, and I admire what you do. But now I like you even more. You're to be commended. You are the walking wounded. The fact that so many of you have been injured and yet remain in such good shape is testimony to your attitude toward physical fitness, personal injury rehabilitation, and your unfailing commitment to this work you seem to love so much."

August 4

Checking our lightning-detection computer yesterday just before lunch, I noticed it was active for the first time in two weeks. While I was standing there, it blipped out strike #253 and gave the position about fifty miles north of Fort Yukon. Minutes later I was on my way to lunch with the first and second loads. The weather had improved and the day was a summer's dream.

When we returned from eating, I went back to the lightning machine. Just before two o'clock, it hit 466 and was averaging three new down strikes per minute. By four it totaled 806, and by six, it had hit 1,146. The first load was sent to Fort Yukon to stand by.

That moved me up to first man, first load. I told a couple snookies, "When down strikes total over eight hundred, and you find yourself on the first load, you better double-check your gear, drain your bladder, and tune your ears for the siren."

Just the same, it had been days since our last fire call, it was dinnertime, and my load was in no mood to listen to Old Leathersack preach about numbers and machines.

"What's for chow?" Kubichek asked, checking the menu on the bulletin board. "Ah! Roast beast. Great! Let's pig out."

Halfway through a large strawberry shortcake, I thought again about being first man, first load. The little green screen had been trying to tell us something all afternoon, and there we sat like fat toads in the moonlight.

"We're the—*burp*—fat first," Erik the Blak said.

"We'd better get back over to the shack," Jack Firestone added, jiggling the van keys at us. "We're gettin' a lot of lightning."

Back at the shack the lightning machine had totaled 1,228 strikes. At 6:49 the siren launched the fat first into action.

"Hey, Oly," I yelled, suiting up at the speed racks. "In the top of the freezer there's a steak. Grab it for me, would you?"

Cramming the hard slab of meat into the left leg pocket of my jump pants, I grabbed my helmet and PG bag and waddled out the door and climbed aboard *Jumpship 19*. We taxied to the east end of the runway. As first man, I sat butt-flat on the floor, with my feet braced on each side of the open door. *Jump 19* didn't have a door that could be opened in flight, so we flew without one. The plane lifted off at a hundred knots, and the ground fell away, 10, 30, 100, 500, 1,500 feet. Eventually the earth drifted far below, leaving me to deal with the feelings evoked by the elegance of an empty sky.

As *Jump 19* leveled off at cruise altitude, I pulled my feet from the sides of the fuselage and placed them more comfortably on the floor. *Jump 19* was packed as tight as a sardine can with six jumpers and a full load of gear.

I thought about Sally and how it was no longer worth looking down for her brother's cabin. I thought about how it had been waking next to Sally the first time, and how since meeting her my life had become both more wonderful and difficult. Since McCarthy, the silence between us had become like a great abyss.

I looked around at the rest of the six-man crew, faceless behind the heavy wire masks of their helmets. I felt a sudden need to make contact. I wanted to know who was who. But it was useless. We were each on our own for the time being, isolated in our own thoughts, hounded by suspense, armored by protective gear, held hostage to an invasive and inescapable roar.

The peaks of the White Mountains—pinnacled, graceful, and arcane—moved by just outside the door, silhouetted against the horizon, some reaching higher than we flew.

At the east end of the Whites, the sky lost its clarity as smoke spread across the Yukon Flats. Farther west the sky lost all definition, and the land disappeared. Nothing was left to look at except the wing of *Jump 19* and a crimson sun moving slowly across the gray rectangle of the open door.

Sore and now cold, my knee began throbbing. We flew into a rain squall. Lightning flashed, heavy turbulence rocked the plane, and the rain hit the wings so hard it sounded like gravel.

Everybody woke up. I was rolling over onto my right side to relieve the pressure on my knee when the plane suddenly fell a hundred feet, hit bottom, and bounced me off the floor. I lay on my side trying to rub my knee, but the pressure of the reserve parachute against my stomach made it hard to breathe. More heavy turbulence. I bounced clear off the floor, slammed down again, and rolled over, the pain bringing me near nausea. Lightning flashed right outside the door so close I thought I could hear it.

We powered on through the storm. Lightning flashed, rain pelted the wings and blew inside, soaking the floor and my legs.

Someone tapped my right foot. It was Kubichek with a note.

Fire B-717—Fort Yukon VOR—27 degrees for 57 miles
Jump 17 already dropped 6 jumpers from Fort Yukon.
5 Acres—Black Spruce—100% active—STEEP hillside.
Fire Boss—Romero —take two radios—Freq. Turquoise
Air Attack—921. Air tankers en route—T-84, T-126, T-33
Helicopters en route—233 the Fort and 363 Central
Go ahead with buddy checks.

Ten minutes later, we cleared the squall and were pulling up on B-717. Dalan Romero was working with *Air Attack 921* and *Tanker 84.* We were directed to orbit the fire at a distance until the airspace was clear for jumping. *Jump 19* found its designated altitude and settled down to 130 knots. The other jumpers bent to the windows, studying the fire.

"You think we'll jump it?" Kubichek asked.

I shrugged. All I knew was that I was ready to get out of the plane. The fire had received a little rain. Pockets of flames were flaring up in the trees along the ridgetop where the tanker was dropping. The rest was only smoke. The rain had flattened the smoke column into a layer of cotton that drifted up-canyon. Air, sweet with the smell of rain and smoke, filled the plane.

A deep draw ran downslope from the fire about three hundred yards to the west; I couldn't tell if it ran water or not. The main ridge ran north-south through the fire and was level on top. The first load had jumped just below the fire. It was steep all right, steeper than the glide angle of our parachutes. Landing parallel to the contour of the slope would be a strict necessity. Too much up the hill and the impact would be head-on, too much downslope and you would drift into the bottom of the canyon. The smoke lay upslope toward the ridgetop— that would be our ground wind. Considering its direction and the contour of the slope, it was clear that we would have to make crosswind landings. The spot wasn't what we normally consider small, but if you miscalculated, the deep canyon, the tall trees, and the steep terrain were interesting penalties to contemplate.

Tanker 84 made low, graceful passes from west to east, barely clearing the ridge, dumping its pink clouds of retardant. I sat in the open door of *Jump 19* and watched. The northern skyline was getting a lot of lightning up in the Chandalar Lake country. Behind the main storm, heavy rain squalls tracked, blending the land and sky together in watercolors of blue and green.

We circled for fifteen to twenty minutes, waiting until the retardant work was finished. Erik the Blak was sitting on some cargo, tightening his leg straps and putting on his PG bag. We started our descent to jump altitude, the other aircraft departed, and Jack Firestone came crawling back through the jumpers with a concerned look on his face.

I tightened my leg straps. We flew south, up-canyon, and Jack dropped the first set of streamers. At times the smoke obscured the jump spot. I watched the streamers twisting and turning in the great space of the canyon. I pulled on my helmet, snapped my chin strap, pulled down my face mask, and secured the collar of my jump jacket tightly under my chin. Jack threw another set of streamers and turned to me.

"Not much drift... Maybe two hundred yards... Stay off the ridge."

He hesitated a second, looked out the door, then turned back.

"Be sure to land sidehill. It's steeper than a cow's face down there."

I nodded.

"Are you ready?"

Kubichek was ready, so I yelled back at Jack. "We're ready!"

"OK, get in the door!"

I slid into the door on my butt and thrust my legs out into the intense pressure of the hundred-knot slipstream. With my left hand over my reserve, I completed my four-point check.

I fixed my eyes on the horizon as the plane leveled out on final. Lightning flashed way off to the north. *Jump 19* eased back on the power. Firestone pulled his head in and lifted his right hand.

"Get ready!"

I grabbed the sides of the door and forced my feet straight out into the wind. I got the slap. I tried to rock forward as hard as I could to miss the edge of the door but bumped it slightly and fell under the tail of *Jump 19* in a slow spin.

I could see Kubichek falling behind me, also spinning. I pulled and opened; he pulled and opened. The soft hum of *Jump 19* retreated into the distance.

The upper-level winds over the jump spot blew one way then the other but mostly down-canyon as the streamers had indicated. Due to the overcast and lateness of the hour, it was getting dark. I set up for a steep final, running up-canyon into the wind. My approach was going as planned, then at six hundred feet I got an unexpected tailwind. I reefed down on my toggles and effected a full stall of the canopy at about four hundred feet to minimize my forward progress. The stall recovery was radical, causing the canopy to fall off behind me, then suddenly pitch forward, leaving me oscillating wildly forward and back. After regaining control, I was still higher than I wanted to be, but now too low to risk another stall. If I continued my present flight, I would overshoot the jump spot and land in the trees. I pulled down on my left toggle and turned straight into the mountain. The jumpers on the ground shouted frantically.

"Sidehill! Sidehill!"

At the last second I initiated a slow off-hand right turn, coming around parallel to the slope and heading right into the jump spot. My orientation to the spot was perfect, except that I'd picked up quite a bit of speed. The tops of trees whizzed by right and left. Suddenly the ground was coming at an alarming rate.

The instant before impact, I noticed a mound of earth the size of a pickup truck right where I was going to hit. I was set up to land side-hill all right but instead plowed into the mound broadside. I had nowhere to go but end over end until my momentum was dissipated into the trees, dirt, and brush, leaving me hanging upside down in some low branches.

"Oh, no," I groaned. Blood came running down the left leg of my jump pants. Both legs hurt. My heart raced. I was instantly nauseous. The jumpers on the ground were yelling, "Are you OK? Are you OK?" I couldn't answer. They were coming. It had finally happened. I was hurt, maybe a broken leg. Whatever it was, I could see that it was serious. My number had, at last, come up, and at my age a broken leg would end my career.

I looked at my leg again, tried moving it. It hurt like hell. Just as the first guys came crashing through the brush, I remembered the steak I'd put in my leg pocket back at the jump shack. It was torn in half and stuck to my pants.

"Hey, buddy, don't move." Buck Nelson said, out of breath.

"That ruined my steak," I told him.

"He's ringy," I heard Dalan Romero say. "He may have been knocked out."

They insisted I stay still until they checked me out. After finding out that the blood belonged to the steak and not me, both rescuers and I felt much better. Once on my feet, both legs seemed to work all right.

Relieved to find out I wasn't injured, the rest of the jumpers went back to the fire, and I was left to investigate the scene of impact. I found pieces of bark peeled off the side of a spruce tree and the other half of my steak. More shreds of bark hung from the face mask of my helmet, but I couldn't recall hitting a tree. I dug six ibuprofen out of the top of my PG bag.

Everyone from our load made the spot in pretty good shape. Clouser overran it and went into the big trees, but didn't hang up. Erik the Blak had drifted off downslope a ways, but soon we were all on the ground laughing and telling stories about our jumps. Before long I was at the top of the ridge running a chain saw, cutting line.

An hour later, Dalan came around to tell us that the fire was in good shape. He suggested that Wally Humphries and I locate a campsite on the ridge and cut a helispot. At the edge of a two-hundred-foot cliff, under a stand of tall dark green spruce we found a small made-to-order flat carpeted with caribou moss. South along the ridge another fifty yards was a small natural clearing that was easily enlarged for a helispot. Once we'd finished the helispot, we went back to the fire and worked to improve the fire line. At 1:00 A.M., with the fire line secured, we headed for camp, me at the end of the line, limping.

From our camp under the tall spruce, the view was west out over the cliff into the big canyon and beyond. The storms had moved farther north in a sky of slate gray, while to the south bands of orange and creamy yellow began to show over the Yukon Flats. With the chain saws at rest, all the world grew peaceful and quiet. Soon a campfire snapped and crackled and we were drinking coffee and heating cans of beef stew. My steaks were peppered with moss and dirt, but they cleaned up nicely. I fried them in the frying pan I carry with me in my leg pocket.

Buck Nelson stepped to the edge of the cliff and gazed out over the canyon. "Tell me," he said. "Are we, or are we not, the greatest fucking heroes that ever lived?"

We all unanimously agreed that certainly we were.

"It's a wonder the babes aren't lined up clawing at the door of the shack just on the slight possibility that they might land a man of such extraordinary mettle."

"We got metal all right," Mitch said. "Plates in our heads and pins in our bones. Don't worry, those Fairbanks women talk to each other. They know what this smokejumping shit means."

"Awk, awk, awk," chortled Erik the Blak, with a laugh that almost hurt our ears. "They know this much," he went on. "When your old man's a jumper, it's best to plan your own summer. Remember when Nemore did that study on marriages between '73 and '89, finding that the average Alaska smokejumper marriage only lasted 2.1 years. Awk, awk, awk."

"Yeah, but you know, those were some busy fire years, too," Mitch said. "In busy years you can almost predict who's not gonna make it till Halloween in the ol' marriage game . . . You know what I mean?"

"I know what you mean," the Blak said. "There's two basic types of jumper wives—three actually—but I'm not counting the ones that stay in the Lower 48. There's the ones who *follow their man* to Fairbanks and don't get a life of their own, so they come over to the shack and hang around watching the jump list all the time. The second kind, you rarely see. They get jobs and friends of their own. When the guy comes home, fine—they can handle it. When he doesn't, they can handle that, too. Strange enough, those are the marriages that last."

"That's too complicated, Blak," Mitch said. "There's only two kinds of women. Those that leave you for others and those that leave others for you."

"Come to think of it, there must be at least four then," Erik the Blak countered, scratching his head. "What about the ones who never leave and the rest who steer clear of men altogether."

"Yeah. And what about that queen in Hawaii that had sixty-five husbands?" Kubi demanded. "That makes five."

"All I'm saying, if I might—just for a moment—interrupt this highly scientific discussion," Buck interjected, "is that if a woman was really serious about wanting a damn good man, she'd set her sights and lay in ambush for a smokejumper."

"I'm goin' to bed," the Blak finally said.

Camp broke and people separated in various directions into the woods. As I hunkered down inside my tent, I heard the Blak working on his hootch and singing a sea lion's version of an old Johnny Cash tune—"I Walk the Line." This from the man who had more experience with marriage than any of us—twenty-five years. When the Blak got to the part about keeping his pants tied up with a piece of twine, some of the guys began chuckling because the Blak had been known to use old ropes for belts. The chuckling had the undesired effect of encouraging him to sing even louder. That went on for a little while, until Buck couldn't stand it anymore. "Blak, Blak. Stop singing," Buck yelled from his tent. "Some grizzly will think there's a wounded moose up here."

"Awk, awk, awk," croaked the Blak.

August 5

At seven o'clock I stirred, checked my watch, then dropped off to sleep again. Waking again at eight, the first thing I noticed was that one side of my tent was dappled with tree shadows. No matter how good it felt lying there in that warm, dry sleeping bag, no matter how tired we were, it was time to get up.

As soon as I moved, I realized I'd been hurt worse than I thought. My left knee was stiff and painful and swollen. Unzipping the bag, I reached for my pants. They were still wet from running around in retardant-covered brush. My boots and socks were soaked, too.

Outside my tent the woods smelled sweet and clean. Small tendrils of steam rose here and there where sunshine hit the sides of logs. There wasn't a hint of wind; all silence and peace. But since Dalan had asked us to be up by eight, I took my frying pan from my PG bag and banged on it with a hard stick.

A few muffled cuss words floated back to me after I had stopped pounding the pan. Then, one after the other, they moaned, stirred, and came tromping out of the woods, red-eyed and scruffy. Within minutes our campfire was going, and the crew was making breakfast as the air filled with the aroma of hot coffee and sizzling flat nose. Dalan took a quick recon of the fire and returned to assure us there was no immediate concern. Hearing the good news, we settled into a leisurely breakfast of cereal, peanut butter, jam, crackers, candy bars, and canned fruit.

"Fire B-717, this is *17 Delta Lima*." It was Dalan's radio.

"This is 717. Go ahead!"

"Good morning. How's everybody doin' down there?" asked Smokey Johnson in the detection ship out of Central.

"We're fine. The fire is looking good—just a couple days of mop-up."

"OK, here's the deal from Central. I'm sure it'll come as no surprise, but they want to pull the jumpers except for yourself and one other. They have a crew en route from Beaver. How on that?"

"We copy."

"The first shuttle should arrive at your fire at 1300 hours, helicopter *38 Juliet*. Can we get all you guys out in two shuttles?"

"Well, there'd be twelve, so I don't see why not."

"OK then, I'll go ahead and notify people on this end. They're expecting more lightning. That's why the rush."

"Copy that."

"OK then, if there's nothing else, we're gonna head outta here and let you guys get back to work. *17 Delta Lima,* clear."

"Looks like they're pushin' the panic button to me," Erik the Blak said. "All this lightning is wet, and it's August. They're gonna spend all that money moving the Beaver crew in here, where if they'd just leave us a couple more days we could finish the whole thing ourselves."

"It would be nice to stay," Dalan agreed. "But it's their call."

"Well," Blak huffed, "they can do it any way they want, but if it was me, I'd hold the Beaver crew on standby in Venetie and wait to see what today's lightning does."

The Blak took out his snuff, tapped the top with his thumb, removed it, and pinched out some Copenhagen.

"Who wants to stay?" Dalan asked.

It would be "choice from the top, boned from the bottom." By jump order, each man could choose to go or stay.

This fire is a good deal, I thought to myself. *I should stay.* But then, when my turn came to speak, "I'll go," I said, thinking of the blood on my jump suit and my sore knee.

After we'd gone completely through the list, Mike Ierien volunteered to stay. Those of us deciding to leave broke down our hootches, ate some lunch, then stretched out on our gear to nap in the sun near the helispot.

The helicopter arrived, circled the fire area a couple times, then lined up on us and began its approach in over the canyon to our spot atop the cliff. We loaded our jump gear, working under the big turning rotors, then climbed aboard and buckled up. As *38 Juliet* lifted off, I looked down and watched Mike as he hunkered near the edge of the helispot, his hair blowing wildly, waving good-bye with one arm, protecting his face from flying debris with the other. Lifting straight up for about thirty feet, then tilting forward, we pounded our way out over the cliff, then dropped down low over the trees and swept out into the great space of the canyon as the fire area below us slipped away.

18

Inactivity infects the crew like a disease. Cooped up in the standby shack, some pace back and forth, bored sick, trying to find something to do. After morning PT, we drift off to various work stations. Some sit in the lounge completing self-study courses in fire behavior or some other training. The placid industry of these days matches that of all the slow days we've had to face before. Present and past are less continuous than synonymous. Day by day the tension builds as jumpers haunt their mailboxes and crab at the box boy over some trivial list change, then walk away, looking lost.

Since the fire north of Venetie we haven't turned a prop. Wednesday afternoon *Jump 17* left for Redmond, Oregon, to finish up the season there. Thursday, *Jump 19* left for Missoula to finish the season with the Forest Service in Region 1.

"It's over!" I hear Seiler say as a group of us gather in the loft. Only I understand that the comment is directed at me. He knows I hate it, and he enjoys getting in a dig when he's feeling playful.

"It's over!" he says again in an annoying fashion, airing his greatest fear—that, in fact, *it is over*.

Our season is evidently going to end early. Word has come from down south that not a lot is happening; the season has been normal, but appears to be cooling rapidly. Unseasonal rains have dampened much of the Northwest.

"It's not over till the fat lady sings," I tell him.

"Yes, but that fat lady has laryngitis," Seiler replies. "She been humped up trying to sing for a week. She hums a little now and then, but she's got the tune down. Her tonsils are starting to quiver."

Even though he's irritating, it's good to see Al healthy again. The effects of his concussion are apparently gone, and his eye surgery has healed without complications. Back on the jump list, Al celebrates by goading me.

"Sounds like you got something going with this fat gal," I say. "You want to tell us about it?"

"You tell us about it. You're the one who brought her up. The only thing I know is that . . . it's over."

Thursday last, five of our jumpers were sent to McCall, Idaho, as part of a prearranged detail. With their departure, the rest of us were left behind to wile away the hours fabricating fantasies about what might or might not happen. Daily we pore over the Boise Interagency Fire Center national situation report.

"It's like playing well the entire season, making it through the play-offs, and then finding out there won't be a Super Bowl," Seiler said. "It's over."

A few of us have begun wearing our heavy down-south boots around the shack to loosen them up. Lighter Alaska-type boots are not allowed by the Forest Service, so our "chicken skins" are switched out for Whites, Wescos, and Nicks. Travel bags have been packed with a few extra goodies that might come in handy in a more civilized setting—a few extra town shirts, your best faded Levi's, a swimsuit, a pair of nice shoes.

Training sessions and special meetings have been held to orient rookies and refresh regulars on the specific hazards of jumping in the Northwest, which overall, is more dangerous than in Alaska. There are a lot of water hazards and some rough terrain in Alaska, but most of the jump country is in the hills and flats of the interior. We have only a few areas of tall trees in the jump country of Alaska, while the Lower 48 is filled with pine, fir, hemlock, tamarack, and cedars that range from 120- to 200-feet tall. The difficulty of rugged terrain, erratic winds, and steering into tall timber were discussed, along with those of heat exhaustion and dehydration. The entire crew has been scheduled for a refresher course in tree climbing. All the while, the "It's over" debate escalates.

"It's over!" Seiler keeps saying.

"We'll be going south, just wait and see," Tyler Robinson insists. "South to America, the land of tall trees and divorcees."

"There's still lots of time," Don Bell snarls. "Remember '87? What happened after August 28? The Siege of '87—that's what! Don't go telling me what's over. I don't want to hear it anymore."

"The siege of your brain," Mitch counters. "What about '83, when we never jumped another fire after August 2? And what about last year? The Lower 48 didn't do shit!"

"Who cares, anyway?" Bell snaps. "We'll just have to wait and see."

"It's over!" Seiler sings just loud enough to be heard across the room.

"It's *not* over, I told you," Bell says.

"It's over, man. Face it!" Mitch teases.

"It's not over, *dammit*! Now don't get me pissed off!"

Mitch turns to Bell. "You know, Don, you need to read some Thoreau. That might help a guy like you. Thoreau talks about people who live lives of quiet desperation . . ."

"Fuck you! You don't know shit," Bell snorts. "And when it comes to jumping, neither does Thoreau, going around preaching to people about how to live. Let him go ahead and live a life of quiet desperation if that's what he likes. I prefer noisy desperation."

"Two things I know!" Mitch says, looking at Bell. "First, I know that it's over, and second, I know that you're an idiot."

Bell launches after Mitch, chasing him around the rigging tables and throwing shot bags at him.

"You'll think it's over." Bell is yelling. "When I get my hands on you, it'll definitely be over."

After lunch I was drafted into the next group for tree-climbing practice. Since there was still a chance we would be going south, it was standard procedure for us to have a climbing refresher. Tree climbing is not one of my favorite activities, but it is an essential smokejumper skill. When parachutes get hung in trees they must be retrieved. Smokejumpers return with everything dropped on their fire. Everything. No matter how big and bad the timber, or how tough the pack out, there are no excuses for not bringing everything out, including the trash.

Back at the shack after practice, I checked my mail and found a letter from Sally.

McCarthy, Alaska
July 24, 1991

Dear Murry,

Hi! Hope you are fine and not working too hard. Just returned from a wonderful flight around some nearby glaciers. The people that own the Lodge are making arrangements for this local who has a Super Cub to fly small groups into remote camps for glacier exploring and hiking. They want to try it for a couple weeks. AND . . . this is the good part: They want me to be their representative at the camps and do the cooking. We'll start around the first of August. I'm having a great time here. So many interesting people from all over the world. We had a beach party two nights ago. We hauled sand from the creek and put it on the volleyball court. Everyone wore bathing suits and pretended we were in Hawaii.

I'm missing you and wish you were here. I'll never forget the time we had. Your gifts were so thoughtful. I love the peach towel and think of you when I hold it close. Being here has me jazzed. I've started doing some art with the stuff you brought down. One is for you. Wonderful! My trip to Alaska is all I could have hoped for. I hope things are going favorably with you. You're probably out on a fire and too busy to think of me. Well, I better get to bed. It's been a long day, and it's back to work tomorrow. Be careful.

Love, Sally

I sat on the bench in front of my locker and read the letter over a couple times. I was thinking of going to see her if things stayed slow, but now that possibility had suddenly closed. No fires, no action, and now with her off in some remote camp, no Sally. I hated to admit it, but I didn't like the sound of her being at a beach party where there were other men her age and with so few women around. The bunch of them cavorting about in bathing suits, flying around, and exploring with adventurer types.

I spent the remainder of the day hanging out in the paracargo bay, trying to distract myself from such thoughts, picking on rookies and snookies, listening to reggae, and watching the rain pour.

August 13

I wrote Sally.

Dear Sally,

It's raining here. My world has come to a grinding halt and yours is in full swing. It's worse than that. I miss you terribly. I wanted to come and see you again, but it sounds like that won't be possible with you away from McCarthy.

I often think of the days I spent there. Such a short time together and so much apart. Lately I don't know what to do with myself. It doesn't look like the Lower 48 is going to have much of a season. Sometimes I wish it would all come to an end, and I could come to McCarthy, take you in my arms, and talk you into a fall trip around Alaska and maybe even down the Alcan through Canada. For now, all I have of you are memories. Sweet, pink, and gray images, and me secretly hoping that sometime we'll be able to share more than a weekend. Having held you close and now having you so far away is difficult.

Take care, and watch out for those bush pilots.

Love, Murry

After dropping the letter in the mail, I hung around the operations desk reading the national situation report. It listed 178 fires for the previous day in the Pacific Northwest. That was about a hundred less than normal. All but three had been contained with in-place forces. Things were not looking good for a trip south.

August 14

After roll call and PT, everyone went their own way. I hung around the operations desk and nursed a lukewarm cup of coffee. On the board next to the jump list, one of our supervisors, sensing the crew's desperation, had written:

IF THERE IS NO WIND, ROW. LATIN PROVERB.

Someone else had written beneath it,

IF THERE ARE NO OARS, DRIFT.

Out on the ramp, air tankers and jump ships sat as they had for the past week—motionless under a gray sky, surrounded by puddles of rain.

I went to the loft and found a parachute to pack. Pulling it from a cart, I threw it on a packing table, hooked the risers to the restraining straps, spread it flat, and began counting packing tabs.

"It's over!" someone sang in a low, needling falsetto. I looked around. Seiler again. Everyone else kept their heads down and continued working. *Everyone wants to be an expert on something,* I thought.

Out of the corner of my eye I saw someone coming through the entryway to the loft. I turned. It was Boats hunched over two crutches, one foot off the floor.

For Boats, Al's nagging little singsong was more than torment, it was fact—at least for this season. Regardless of what happened down south, Boats wouldn't be part of it. All he had to look forward to was a winter of rehabilitation work and a lot of road miles building his knee back up again.

"Hey, Boats. How's it goin', dude?" Dunning asked. A group gathered around.

"So far the Doc says it went well. I guess they got what they went after."

"The meniscus?"

"Yeah! But not real bad. More like a typical initial surgery—a few minor tears is all."

Boats set his crutches aside, turned around, and lifted himself gently onto a packing table. Sliding back, he lifted his leg. The pant leg had been cut lengthwise. Boats undid the safety pins and unwrapped the bandages. Then the knee—black-and-blue, red and yellow, swollen and shiny. The surgery entry points looked like a large snake bite—two black holes, one on each side of the kneecap.

"They went in here," he said, pointing to the holes, "and cut away the damaged part of the meniscus. It's not that big a deal, really."

I went back to packing the chute and listening to Gene Autry. Gene was riding Champion out through the desert somewhere, with not a care in the world, singing "Back in the saddle again."

I've never been seriously injured, but I know that once you have, things change. Gone is the illusion of invincibility. No one believes it can happen to them—until it does. You find yourself hobbling around the shack swapping stories about broken body parts, your name tag pulled out of the lineup and put in the lower right corner of the jump list under "cripples." A heightened sense of vulnerability may mark the onset of fears that can gradually erode a jumper's confidence, crippling his spirit along with his body.

Most keep the faith, though. Dave Mellin broke his back three times before he quit. And Floyd Whitaker told me once, "I wanted to jump again as soon as the doctors released me. I had to go back up and do it because I knew I had to show myself I could." The summer before, Floyd had miraculously survived a total streamer of both his main and reserve parachutes. On a fire jump out of Missoula, he fell fifteen hundred feet to crash down through the side of a tree, breaking branches, ripping his chutes, and then slam into the hillside and skid forty feet down an exceptionally steep slope. The jumpers on the ground heard Floyd scream as he disappeared into the timber. His twin brother, Lloyd, was later heard to say, "Hell, it would've killed anybody else."

Floyd had broken his leg, his back, an arm, and "a few other things." When the other jumpers got to him, he was sitting on a log puffing on a Marlboro.

"I don't remember hitting," he said, "but when I came to I jumped up and ran around in a little circle, then fell right over 'cause my leg was broke."

Floyd returned to jumping and worked for several more years. I saw him in Missoula in 1988 while Yellowstone was burning. After more than twenty years of jumping, he'd accumulated several injuries. Always the old ones kept coming back to bother him. When the doctors warned him that he risked becoming paralyzed if he hurt his back again, he finally decided to call it quits.

"I loved jumpin'," he said, grinning, a silver-haired old badger. "But, I like walking, too."

August 22

The "it's over" comment is no longer funny or even debatable. Most agree, if it's not over, it's too close to quibble about. We've had a good season for sure, but one that has ended too soon.

For a few of the jumpers, it's definitely over. Kubichek caught the red-eye special to Seattle last night at midnight. There was a round of handshakes down at Pike's, and then Kubi left for the airport about eleven o'clock. Kubi had turned in a very successful season as a snookie and was turning out to be a fine smokejumper. We had worked a lot of fires together, had been jump partners several times, and I was sad to see him go. He was going to Missoula to study for his commercial pilot's license. Rick Russell and Chris Woods left at six last night, right after work. They left without a word, California bound in Rick's old Volvo with his green canoe strapped upside down on top. Paul Sulinski has also gone—headed for San Francisco—then on to Japan and Tibet.

Open spaces have appeared on the suit-up racks for the first time in three months. There's a different sound in the shack, a different look to the jump list. Several names have been moved to the lower right-hand corner and turned upside down. The airport is peaceful and quiet.

Any chance of work down south slips further and further away with the passing of each day. With a slightly higher than average number of starts from lightning, plus the normal man-caused frequency, the Northwest had remained promising until recent rains. From the Interagency Fire Center in Boise, word has come that "Due to wet weather and fall-like temperatures, there will be no need for Alaska smokejumpers for the remainder of the season."

After work I went to my room in the barracks and lay on the bed trying to figure a way out of the funk I was in. Most of the jumpers on the crew were handling the downtime better than I was. My inner tendency toward melancholy has never fared well having to deal with boredom.

There was a knock at the door. "Come in," I said. It was Bell.

"Hey, Don, what's up? Grab a chair."

"Yeah," he said, shuffling his feet and scratching the back of his head and fidgeting around self-consciously.

I said, "You must be looking for those two rollies that just cruised by here headed for Fairbanks."

Don's body jerked, then went rigid. "What? You saw two rollies headed for . . . Naw . . . Ha, ha. There aren't no rollies going anywhere.

Me and Big Mac got all the rollagons locked up over in the fuel yard inside a chain-link fence."

"Want a beer?"

"Oh, sure. You got one?"

I went to the refrigerator in the corner of my room and took out two bottles. I twisted the tops off and handed one to Don. He lifted his in my direction.

"Here's to a good season," he said. "We had a time, didn't we?"

"That we did," I said. "Good working with you."

"I guess you've heard I'll be leaving tonight. In just about an hour," Don said. "I've got to get out of here. I'll do something bad if I stay any longer. This slow shit's driving me nuts, along with everybody else. I don't have much to go to, but I'm going anyway."

"To Tillamook?"

"Oh, no," Don answered, surprised. "I'm not leaving Alaska. I'm going into the heart of it. Fall's my favorite time up here. I'm packing in to my land and spending a month hiking around and figuring out how to build a cabin."

We talked a little longer about his land and how exciting it was to commit to building a cabin. By the time he'd finished his beer, he was looking nervous again.

The calling of a group of Canada geese came to us through the window. Passing by a hundred yards north of the barracks, they flew east up the river.

"Hear that?" Don asked excited. "They're gathering up. I saw some yesterday at the end of the airfield bunched up like they do before they head south."

The geese moved up the Chena River, their cries fading away.

"Once I hear them, that does it for me. I've got to get moving."

Don stood up and went to the window and looked off in the direction the geese had gone.

"I wanted to come by and see you before I left, though. I know you've been having a hard time lately. You and your woman troubles. I'm no expert on the subject, but I'll talk about it if you want to."

I told him how the summer had gone, me always going one way and Sally the other. I explained that now that I had time on my hands she had gone off to some remote camp, and so it wouldn't do any good to go to McCarthy.

Don listened, still looking out the window, fidgeting his feet around, furrowing his brow, impatient for me to finish my story.

"Look," he suddenly interrupted. "You want to see her. *Go see her!* She's down there somewhere. Find her. Why hang around here feeling sorry for yourself?"

"Just go down there?"

"Why not? Fire season's over. Charter a bush plane and fly to her camp. Take a gear bag and pack in to see her. The worst that could happen is you'd have a great hike. Anyway, that's what I come to tell you. Now I've got to go on."

Don turned from the window and held out his hand. "Maybe I'll drop by and see you sometime this winter," he said.

We shook hands and he left. I began to shift mental gears. I was only making straight time, eight hours a day. I could take a week off, drive to McCarthy, and have the guy with the Super Cub fly me in just like any other paying customer. If that didn't work, then I'd walk the thirty or so miles to her camp just like Don said. What an idea! Sally would look up from her cooking fire, and there I'd be, walking out of the bush and into her camp. It was perfect. The more I thought about it, the more I knew I was going to do it.

August 25

Beep-beep-beep-beep. I rolled over reaching for the frantic little clock and knocked it off the dresser and halfway across the room. I switched on the light and sat heavily down on the edge of the bed. The clock blinked 4:01. I looked back at the bed and thought about Sally.

I couldn't believe it. I'd spent two days getting ready for my trip to McCarthy; another shopping trip for her, bought a bunch of extra camping food, filled out leave slips. I was filled with the anticipation of getting out of Fairbanks and once again holding Sally in my arms. This time I would stay as long as I wanted. I'd planned to leave yesterday at noon but then all that changed.

I pulled on my Levi's, stepped into the bathroom, splashed water on my face, dressed, grabbed my travel bag, turned off the light, and closed the door quietly behind me. Outside, the morning air was warm and intimate and smelled of fall. At the standby shack, I locked my pickup and went inside to join a group of sleepy eyes gathered around the box talking softly.

"We've been to the play-offs," Al Seiler said. "Now it's time for the Super Bowl." I shot him a hard look. "Hah! And all the time you thought it was over," he said, grinning.

On the jump list written in red felt-tip pen I read what we already knew:

20 JUMPERS TO MCCALL—ASAP—1ST GROUP 11:00 P.M. TONIGHT
—2ND GROUP 5:00 A.M. TOMORROW

Mitch was there. Togie, Tyler, Troop, Buck, Olson, Chip, and Quacks. Firestone was our COP (Chief of Party). Of the twenty requested, I was the last to get to go. The Northwest had had three days of dry lightning, hundreds of new fires, and most bases were low on jumpers. Gear bags, travel bags, and fire-pack boxes filled with extra parachutes were loaded in the paracargo van just outside the receiving bay of the parachute loft. Once everything was securely on board, we piled in the blue van and left for Fairbanks International.

"Boys, we're goin' south, to a land of steep mountains, starry nights, hayfields, beautiful women, and fires," Chip Houde said. "You gotta like that."

"I can't believe it's real," Togie said. "I'd lost hope."

"That's the thing," Mitch said, turning and looking back toward the rear of the van. "With this job, you never know. But once they get us down there, they'll keep us for a while, no matter what."

Inside the van, the group calmed down under the cozy spell of the heater, dash lights, and soft country music. The blue lights of the runway glided past as I whispered a private good-bye to Sally, to Alaska, and added a little prayer that we'd all make it back safe and sound.

Out on Airport Road, Jim Olson suddenly spoke up.

"OK, you guys. What should we be thinking about jumping down south?"

"Tall trees and divorcees," Tyler said, looking out into the darkness.

Oly turned around bug-eyed. "Tall trees and your ass hung up in one," Oly said.

"Best review your letdown procedures," Buck Nelson said.

"Think about it," Oly added. "Before the day's over you might find yourself hanging in the top of a 150-foot Doug fir, and you can be damn sure there won't be any divorcees around to help you get down."

At the main terminal we unloaded three thousand pounds of gear and carried it all inside. Regular passengers looked on in amazement at our mountain of equipment.

After completing our check-in, Firestone distributed the tickets and boarding passes and we headed for the coffee shop for a forty-five-minute wait. It was 6:30, and I was feeling drowsy. Over coffee we bullshitted about other times down south.

"In '88," I said, "I jumped thirteen fires and worked fifty-six straight days."

"Yeah," Seiler groaned. "I was never so trashed in all my life. Remember that flight to Missoula from Great Falls? That was the day the lodge at Old Faithful nearly burned, and the canyon fire came down out of the mountains and burned across miles and miles of wheat fields. Tractors were tearing around in circles, plowing lines around houses and barns. I thought the whole world was going to burn."

"'88 was epic," Mitch said. "I logged 455 hours of overtime in forty-eight days. It won't be anything like that this year, but at least we're out of here."

At 6:40 Oly walked up to our table with a stricken look on his face.

"We've been canceled!" he said matter-of-factly.

"What?"

"No shit," Firestone said, walking up behind Oly. "I just got a call from Jones up in dispatch. Boise called a few minutes ago and canceled us."

To say that it was a solemn bunch that rode in the blue van back to the shack that early dawn would be an understatement. The feeling was one of utter disappointment, disgust—an impossible irritation. Never had we been that close and been canceled. Of all the things that could have happened, not one of us was ready for anything that depressing.

"Fuck BIFC," Quacks said. "Those weaktits. What the hell's going on with this outfit anyway? They jack us around like pawns in some weird little board game."

"I knew it," Firestone said. "We should have gone last night with the others."

"Fuck it!" Buck said in a manner much out of character. "We're gettin' boned!"

"That's OK, fellas," Troop said. "We'll get our chance again, someday."

"This is the biggest bone job in history," Mitch grumbled. "It's a good thing we reviewed our letdown procedures. This is a fucking *letdown,* if I ever saw one."

"I can't believe it," Al kept repeating to himself. "I just can't believe it."

The Interagency Fire Center had canceled us without explanation. There was nothing we could do about it but bitch and bellyache. We pulled up to the bay doors of the parachute loft and unloaded our gear. After filling out a time slip, I went back to my room and unpacked my travel bag and fell into bed.

Around four that afternoon I woke to the pale light of a fading day and a gray overcast. Wind blew through the open windows above the main stairwell in the barracks and set up a dreadful moaning sound that traveled up and down the halls, a sound that resonated with something in me that just wanted to start tearing the doors off their hinges and kicking holes in the walls. I got up off the bed and started pacing back and forth. I'd had enough loss—for Sally, for the other jumpers quitting—and now with the canceling of our trip, it was all just too fucking much.

It *was* over. It was late August and the thought of another month hanging out at Fort Wainwright, working straight eights, and going nuts at the same time was grim indeed.

"To hell with this bullshit," I muttered.

I needed to sleep. Then I would get the hell out of that cuckoo's nest. I lay down on the bed, buried my head under the pillow, thrashed around a while, then finally fell asleep.

Just before seven I woke up as the rain and wind beat against my window.

There came a rapping at the door. "Hey, Taylor." It was Seiler.

"Yeah?"

"Hey, man. You still want to go south?"

"South?" I mumbled. "Hey, Al, fuck you. I'm in no mood for jokes."

"Down south, buddy! Boise called back. We're on the road again."

Riding to the airport the rest of the jumpers shook their heads, partly disgusted and partly amused by the absurdity of it all. I sat quietly in the back of the van convinced I was losing my mind. Come, go, stay. Wrung out. Too many loose ends. Whatever our fate was, I needed to get on with it. Ten minutes before midnight we boarded the red-eye and took off, powering up through the clouds and the rain until we broke out on top. As we leveled off somewhere near forty thousand feet, I stared out the window at a waxing crescent moon with Venus sitting bright and serene right beside it. The earth had disappeared far below, leaving its curvature outlined by a solitary yellow-green band of light that spread across the northern horizon, separating our planet from the deep blackness of space.

I wrote Sally a letter.

August 26, half past midnight

Dear Sally,

I have to write you. Things have been crazy. In a few minutes I'll
be flying over your little camp somewhere down there. Up here at
forty thousand feet, all I can see is a new moon and a few old stars.
I've missed you much these days. My life has been strange. I was
coming to see you. I hadn't figured out how yet, but I was coming
just the same. First I was coming, then we were leaving to go south.
That was yesterday. Then we were canceled. Now we're going
again—to McCall, Idaho. I just wanted to write to you as I'm flying
over. It's the closest I've been to you since I was there. I suppose
you're sound asleep down there somewhere in that great darkness.

Sweet dreams, Murry

August 26

Sitting next to the window of the De Havilland Dash 8, I marveled
at the hazy blue silhouette of Mount Rainier, then Mount Saint Helens
with eighteen hundred feet of its top blown away. To the east, the sun
rose above a gray band of clouds. Below I could see the gaping mouth
of the crater. Steam rolled up and turned into gold vapor as it cleared
the rim. Landing soon at Portland, we would catch another plane to
Boise, then still another on to McCall.

I glanced over at Quacks, who was sitting next to me. Quacks had
ended last season crashing out of a ponderosa pine in eastern Oregon,
dislocating a shoulder and cracking some ribs. Mount Hood, Mount
Jefferson, and the Three Sisters stood to the southeast, where a line of
cumulus clouds was already building over the mountains.

"The scouts are out," I said, pointing out clouds that forewarn of
lightning.

"Man," Quacks said. "This is bad-ass down here. Just look at
those mountains. You mess up up north and you get shit from the bros.
You mess up down here, and you're going to the hospital."

We arrived in Boise at eleven that morning, picked up our gear at the baggage area, and waited for the vans coming down from McCall to pick us up. It was hot. We spread out on the lawn, resting our heads on travel bags and trying to sleep as passengers scurried by on the sidewalks.

At 12:30 a woman in a Forest Service uniform arrived from BIFC and said our plans had changed. We would be flying to McCall instead of driving. McCall was jumped out, and a plane was standing by for us if we got a fire call before dark.

At two that afternoon we took off, flying north over the Boise National Forest and on into the Payette. Twin Otter *40 Zulu* bounced and bucked through the turbulent air while thunder cells towered all around us. Some were able to sleep; others just sat quietly, eyes closed, resting as best they could. The Boise front country gave way to the higher mountains, heavily timbered ridges, and deep canyons. Then, up ahead, Cascade Valley appeared rich and green—McCall and Payette Lake.

The valley is a patchwork of small farms, green and yellow fields, streams, and country roads. To the north, along the south shore of Payette Lake, lies McCall—the quintessential alpine summer resort, complete with sky blue waters, a small sailboat marina, Boy Scout camps, and smatterings of Old West architecture. Surrounding the valley, the lake, and the town, extending in all directions for as far as you can see, are the great wilderness mountains of central Idaho.

The McCall smokejumper base is surrounded by hayfields, the Payette River, sagebrush, and lofty stands of ponderosa pine. Minutes after our arrival and a few handshakes, McCall operations foreman John Humphries called us upstairs for a briefing.

"Just a few things before we get you guys going," he said, hesitating a moment to check his notes. "Anybody not been here before?"

Togie held up his hand. Humphries smiled warmly.

John is a tall, broad-shouldered, softspoken, sincere man with straight blond hair.

"Sorry about the mix-up in getting you guys down here," he said. "Somehow BIFC got their wires crossed. We've been pretty busy this last week. Actually, we've been fairly busy all summer. Nothing bust-like or overwhelming, just steady, the way we like it. Then about a week ago we were jumped out a couple days in a row. We did that yesterday and today, too. Our guys are getting back pretty quick, but the districts are getting too many fires and the turn-around times are getting longer. Plus, the guys are starting to show the wear and tear."

John hesitated and smiled again, his management style that of a person whose priority is people, first, second, and always. Amiable and easygoing, he enjoys a good laugh, the satisfaction of his work, and seeing old friends again like Davy Hade and Jack Firestone, who used to jump there.

"Most of the action has been over on the Nez Perce and here on the Payette . . . also some on the Challis, the Sawtooth, and a few on the Boise. Now the wildernesses up north have taken a pounding."

I thought about the wildernesses of which John spoke. Idaho license plates read "Famous Potatoes," but they should read "Wilderness Forever." When it comes to officially designated wilderness, Idaho ranks number one in the nation: the Gospel Hump, the Pioneer, the Clearwater, the Selway-Bitterroot, the Seven Devils, and the River of No Return, to name a few. Within these remote strongholds, names like Chamberlain Basin, Moose Creek, Magruder, Lochsa, the Salmon River breaks, and White Bird all stir in smokejumpers the yearning to set foot there again.

Completing the orientation, John wadded up his notes.

"So! It's good to have you guys. Hope you have a good stay. We dropped the first bunch you sent down this morning, and we've got one load here now. They got back in just before you showed up. There's more due in by seven tonight."

John hesitated a moment, took aim, then tossed his wadded-up notes into a distant waste can.

"I know you guys are bushed. But as long as we can handle the fire calls, we'd like to get in a practice jump and some tree climbing before dark. You'll be out of here first thing in the morning and into the big trees, so this will just be a little refresher."

The turbine DC-3 lifted smoothly off the north end of the McCall runway and passed over town, continuing its climb out over Payette Lake. Shortly after leveling off, we began to hit rough air. Our spotter pulled on his goggles, put on his headset, picked up a fistful of streamers and lay belly down on the floor, thrusting his head out into the slipstream.

I looked down at the jump spot. It was one of their usual practice spots and a good place to climb trees. There were sufficient openings between the trees to allow a proficient chute handler to avoid hanging up, but god help you if you did, for they were all monstrously tall: ponderosa pine, sugar pine, and, of course, towering Douglas fir.

The air had been turbulent the day before, too, and the ten Alaska jumpers that came before us had not impressed the McCall ground crew. Blowing all over the place in their square parachutes, some jumpers missed the spot by as much as two hundred yards. A couple nearly wiped out, brushing the sides of big trees and partially collapsing their canopies. Charlie Brown tumbled end over end as his chute dragged along, snagging branches and tearing. Thirty feet above the ground it finally caught on a limb strong enough to stop his fall. A grateful and shaken Charlie had to do a short letdown.

We would be jumping in three-man sticks—I was third man, first stick.

"Two hundred yards of drift. The wind's a bit tricky down there, mostly out of the south," spotter Barry Koncinski shouted over the roar of the open door and the left engine, only fifteen feet away. "Did you see the streamers?"

The first two jumpers nodded. I did, too. Although I hadn't seen them much, I got the general idea—the winds were shifting back and forth, southeast to southwest, and around and around. That would be tricky, all right. If the jump spot was not bad, the air we were flying in certainly was. Bob Quillin once described smokejumping as "prolonged periods of boredom punctuated by moments of stark terror." I was thinking about that as I hooked up, with the Doug slamming hard, up and down, tipping side to side.

"Get in the door," the spotter yelled, and the first man in our stick did. That was Troop. Mitch was second. I was last and thus would be carried the farthest upwind.

Just seconds from our exit point, we hit a pocket of down air and dropped a hundred feet, then stopped abruptly, causing my knees almost to buckle.

"Get ready!"

It was a routine get-out. I watched the big broad tail pass over my head and pull away as I went through my count. The opening was normal. I'd packed it myself, so the pretty parachute overhead was doubly satisfying. I quickly checked my canopy, checked my three rings, disconnected the Stevens, and reached for my steering toggles. Suddenly I dropped as if I had cut away from the chute. The quick drop ended in a violent jerk, and I looked up. The canopy seemed fine except that the left end cell had closed under the skirt, and the canopy rocked crazily back and forth.

For the next ten minutes I fought some of the most radical air I've ever been blown around in. The wind kept switching north to south, then north again. I held above the spot, upwind to the north. Mitch had gone low and south. Troop was yo-yoing up and down. Direction wasn't the problem. The square I was flying could counter winds up to twenty-five miles per hour, but radical up air kept lifting me back into the sky. I stayed upwind and rode the air currents, but I wasn't descending. Pulling the front risers down under my chin, I tried to plane, but that was a lot of work and accomplished little. It took a lot of strength, and for thirty seconds planing, the up air would lift me back up in ten. Up air like that was unusual—even for the Lower 48. At times it felt like I was tied to a wild horse instead of a parachute. Down below I noticed Mitch, circling like a buzzard, trying to figure out his approach into the spot.

I pulled down my front right riser and did a series of bomb turns that stood the canopy on its edge and whipped me around like a ball on the end of a tether. Falling at eighty feet per rotation, I was getting dizzy, but I was also losing a lot of altitude. I stayed with it, dropped about a thousand feet and got below the stronger thermals. At about five hundred feet, I set up my final from the northeast. I didn't see Mitch. Troop I had left way up high.

There was an alleyway from the north into the spot, and there seemed to be a slight head wind. The wind sock on the spot wasn't telling me anything, first blowing one way, then the other. At 250 feet I was fully committed to my approach when an updraft caught me face on and began lifting me, ever so easily, back up away from the spot. I held hard into the wind, not daring to turn aside or downwind into a stand of 150-foot trees.

Like a kite, I soared steadily back up to about three hundred feet, drifting north and east. Troop would need room—I had to get down out of his way. The next thing I knew, the wind let up and my chute took me back in the direction of the spot again. At treetop level, the same thing happened again, and I climbed back up, drifting within fifteen feet of the lofty tops of a couple of enormous ponderosas. Time was running out. I held back to the east and attempted to enter the alleyway again. The wind sock still showed the ground winds as variable. The wind subsided. I began to move forward and soon was flying in between trees, still a hundred feet up but descending steadily.

Just when I thought all was going well, I found myself accelerating forward on a tailwind. I sailed over the spot as if in a wind tunnel, passed the Forest Service trucks, and glided on across the road, turning quick left, quick right, then quick left to avoid brushing the sides of trees, and finally piled into the ground, barely missing a big granite boulder, but skidding into a huge pile of fresh cow shit.

Too far from the spot to feel good about anything, I was airsick, embarrassed as hell, and disgusted with the entire mess. Old Leathersack, the old salt, had missed the spot and got pasted with cow shit to boot. Damn the luck! I hadn't had that close a look at a pasture pizza since my early years at Redding when we used to practice jump out near Palo Cedro.

Hell, I thought, *be glad you're alive. You're not hurt, and you're not up in one of these trees thrashing around with your letdown rope.* I looked over toward the jump spot, and the first thing I saw was Togie Wiehl coming down perfectly, nearly hitting the X. His first jump in the Lower 48. His first in big timber. Coming in like a feather for a stand-up landing. Rookies. Sometimes I think the world would be better off without them.

I felt weak and smelled like cow shit as we gathered around Humphries for our brief refresher on tree climbing. We sat in a semicircle under a big sugar pine, fingering fallen pine needles, while John leaned against it and spoke from his notes.

After that, we spread out to different trees and did some climbing up to about sixty feet and performed a couple limb-overs. Back on the ground, I talked casually with John while the others stepped up the broad trunks, throwing the big ropes, then stepping again, while flakes of bark sprinkled down through the afternoon haze and fell quietly around us.

After the tree-climbing session we returned to the base and unpacked our jump gear. I went outside and hosed the cow shit off mine and then arranged it on a fence to dry. I'd had enough. In the past sixty hours, other than a few catnaps, I'd only slept eight hours. In the previous twenty-four, we'd flown thirty-five hundred miles in five different airplanes, had a wild parachute ride, and a workout climbing trees. My guts burned from too much coffee, my head ached, and my knee hurt, but at least we had made it down south.

"We're gonna have to put you guys up tonight at the forestry camp," Rick Hudson explained from behind the ops desk. "All the hotels and motels are booked up for the Labor Day weekend. It's nice though; you'll like it. It's by the lake."

The forestry camp turned out to be the summer camp for the University of Idaho Forestry Department. The sleeping quarters were log cabins nestled under massive old-growth pines, six people to a room, bunk beds and sleeping bags. The prospect of spending the night lying awake in the dark with a roomful of snoring, farting smoke-jumpers was the last straw. I grabbed my mattress and sleeping bag and walked down a sandy beach to a floating dock. At the very end, I placed the mattress right next to the water, spread out my sleeping bag, and lay down.

Majestic pines lined the lakeshore. Scott Dewitz, Togie, and Buck came down and took a swim. The western horizon paled yellow under a wash of blues and greens. A fingernail-thin crescent moon appeared. Timber along the far shoreline formed into a dark, jagged line that divided the sky from its reflection in the lake. A few stars came out. Meanwhile the swimmers had returned to their cabins, leaving me to rock gently on a bed surrounded by water, while small waves lapped at the shore and the lights from town came slithering across the lake like neon snakes.

August 27

The Forest Service van turned west off the highway and crossed perpendicular to the north end of the airport.

"The scouts are out," Troop said.

To the southeast, beyond Cascade and Long Valley, a string of early morning cumulus clouds had formed along the crest of the Sawtooths.

"Cumulus overtimus," Mitch said. "Money in the pocket, shoes for the kids."

"We're gonna see some smoke today," Seiler said, waving a finger out his driver's-side window toward the clouds. "Better make sure all your gear's ready cause we be doin' some flyin'."

The farther out in the valley we drove, the more we could see the buildups. It was a typical late-August morning in the high country of the Northwest. A southwesterly flow of moisture-laden marine air was in off the Pacific, moving across northern California, eastern Oregon, and on into central Idaho, Wyoming, and Montana. The warm, wet air, lifted by the mountains, cools, condenses, and forms clouds. By early afternoon the clouds would tower to twenty-five thousand feet and cast great shadows upon the forests, farms, and fields. By late afternoon static electricity created by the turbulent inner workings of these clouds would suddenly arc, cloud to cloud, cloud to ground.

When lightning hits, trees explode, split in half, and scatter in flying chunks. Shafts of fire dance about the base of trees, igniting leaves, twigs, and rotten logs. Smoke appears—small at first, then with flame. Finding its way to more fuel, the fire makes a small run. Smoke filters up through the forest canopy, white, blue, and pungent. Sometime later, maybe that day, the next, or even next week, a lookout will aim binoculars, make quick notes, and reach for the radio.

Walking into the ready room, we found some McCall jumpers suiting up.

"Hey, you guys are just in time," a cheerful John Humphries said from the box. "Got a fire call—a holdover in the Seven Devils. Some of you guys will make the load."

John turned and ran his finger down the jump list.

"Baumgartner, Houde, Wiehl, Nelson . . . it's going to be the Doug, so get your shit ready. The pilots should be here any minute."

Alaska jumpers had been interspersed down through the McCall list. Some other McCall jumpers had come in during the night, so after they rearranged and interspersed the Alaska bunch, I found myself about the middle of the fourth load. The first load suited up and took off west for the Seven Devils. Within the hour another load left, and then things went quiet. At noon, when we went to lunch, I was still somewhere on the third load. By five that afternoon another Doug load had flown north across the South Fork of the Salmon and dropped all twelve on a fire in the River of No Return Wilderness. That made me fifth man, first load. For the next hour I sat on a folding chair in the ready room inspecting deployment bags, feeling uneasy.

Twin Otter *412* returned with four jumpers still on board after dropping six in the Imnaha River gorge in eastern Oregon. That moved me down the jump list to first man, second load. Those of us still there wandered outside, kicked back on the lawn, told jump stories, and tried not to think about what might happen next.

The jump stories got around to the subject of delayed openings, and Tyler wanted to hear the story of Freefall Hall. Freefall Hall never actually had a delayed opening—he almost had none at all. Freefall, as he was later called, was a man of artistic leanings and liked to talk of photography, art, foreign lands, or anything else for that matter. Be it an in-depth critique of early Ansel Adams or a passing note from Helmut Diller's *Field Guide to the Mammals of Africa—including Madagascar,* Freefall might just as readily report on the importance of dramatic contrast in black-and-white photography as he would on the natural history of the giant Gambian rat, an inhabitant of the Congo River Basin that is remarkable for its habit of taking a shit while doing a handstand.

Back in 1972, while circling over a fire in the Innoko Flats of Alaska, Freefall had become engrossed in some such topic with the jumper sitting next to him. When it came his turn to jump, he stood up, pulled on his helmet, stepped up behind the first man in the door, and went right on talking. With the plane on final and his jump partner firmly planted in the door, he was hurrying to wrap up his story when the spotter noticed that his static line was not hooked to the static line cable but was still tucked under the emergency knife on his reserve. The spotter shot his hand up blocking the door and the plane did another go-around while Freefall hooked up properly. From then on he was known as Freefall Hall.

At 6:30 the question of what would happen next was resolved for everyone but Troop, Quacks, and me. The siren brought us up off the grass to hit the turf running. Within minutes, the Twin Otter was running up its engines as the three of us stood by to see the rest of them off. They waddled out to the plane looking absurdly serious, the bulgy asses of their jumpsuits giving some of them what we call rudder butt.

"Have a good one," I yelled.

That moved me up to first man, first load. I went in to make another check on my gear and to review what Humphries had told us about local radio procedures. The base had been cleaned out except for the last load of seven. Troop and Quacks and I were the only Alaska jumpers left. Rumor had it that Missoula had had another jumped-out day—same for Grangeville. In the next hour, the lightning continued to flash in the east country, but the siren stayed quiet. I expected it at any minute, so every little sound—the start of the air conditioner, the squeak of a door, the horn of a distant car—had me jumpy.

The sun dropped down below the clouds and burned clear just above the horizon. For a few minutes the valley glowed with a vivid amber light against the darkness of the east. Clouds gathered in pinks overhead. The evening air began to cool, and at 7:45 it came over the PA system that if we didn't get a fire call in fifteen minutes we would all be off at eight.

And so a long day of standby and nervous tension came to a close. A day that I'd spent sure that I'd be jumping had ended anticlimactically with a set of strung-out nerves and a ride to town in a Forest Service van.

Later that night a thin moon appeared vaguely through the tattered edges of clouds as I stood on the back deck of the yacht club looking out over Payette Lake. I could see the faint shoreline timber and the dark mountains beyond. Somewhere out there, far into the wilderness, my smokejumping brothers were scratching fire line, falling snags, and making plans to catch their fires before morning.

I stood for a moment surveying the small marina. Behind me, music blared from the open windows and doors of the yacht club, then dissipated weakly out over the lake. I thought about Sally.

Sometimes it all seems so confusing. Maybe that's the problem with wanting too much out of life. Wanting badly to see her, I had decided to go there as Don had suggested. Then when the call came to go south, I had chosen jumping fires over her. Now I was thirty-five hundred miles from where she was, and choice was no longer part of the deal. My mindset was fires, and there was no way to change that so quickly. Even though Alaska had turned cold and rainy, our heads and our hearts still hoped for the down south part of summer. Now here I was, standing alone in a beautiful place, thinking about what I'd left behind.

Back inside I took a seat on a barstool and ordered a whiskey. The jumpers on the last load and a few others who had just come in off fires were in a partying mood. I felt like dancing, so I asked. My partner liked to dance and smiled a lot, so we danced three in a row. Dark hair, large brown eyes, a few freckles, and light skin. She looked to be in her late twenties and was packed into her white satin cowboy shirt and black Levi's as tight as a reserve parachute.

The band took a long break. When the music started again, I thought about asking for another dance but decided not to. I'd spent enough nights aching for a woman, involving myself in all kinds of roguish nonsense trying to impress one, and then having to leave empty-handed anyway. I was first man, first load. *One more drink and I'm out of here,* I told myself.

I felt a tug on my arm. It was my dancing friend, and she wanted to dance some more. We went out on the floor and went at it. After a couple of fast dances, we danced a slow one. She grabbed me tight, and I could feel her against my chest as we danced cheek to cheek. At the end of the song she turned her head, presenting her lips. Without thinking, I kissed them.

We left that place on fire. The woman I had just met led me by the hand down Main Street, then left up the hill. There were few street-lights, so I had no idea where I was going. Her name was Carol. Every so often we'd stop, grab each other, and waltz around in the street like Russian bears. We came to a corner lit by a streetlight where a little dirt road led to a tiny cottage under some big pines.

She reached above the door and took down the key.

"Don't mind the dogs," Carol said. "They're just friendly."

As she opened the door two enormous dogs, an Airedale-looking monster and some other experiment in big hair, exploded in a frenzy of jumping and pawing, knocking over kitchen chairs, whining, and nosing me in the crotch. They jumped at Carol and ran in circles barking. The noise in the little house was deafening. What a relief when they finally flew out the door and into the night. Carol slammed the door after them.

"You'll have to excuse the house, too," she said. "I've been working double shifts."

We kissed right in the middle of the tiny kitchen. Carol made a satisfying moan. Unbuttoning her blouse, I slipped my hand over her satin bra. Her breasts were large and bulged around deep cleavage. I rubbed my cheeks along her ample contours as she reached up to undo the center of the bra. In the faint light of the streetlight, I nuzzled and kissed her. She pulled out my shirttail and unbuttoned my shirt. She tugged her bra aside and we pressed together. I reached around her as she stood in front of me, and pulled her tight. She ran her hands through my hair, and I began unbuttoning her Levi's.

Suddenly there was a terrible racket just outside the door, a dogfight. Carol yelled. "Kelly! Gretch! Knock it off! Bad dogs!"

The scolding initiated more barking and an outburst of scratching at the door. "Shit," Carol hissed, turning to let them in. They sprang into the room like kangaroos, slobbering, whining, jumping on us. Carol cursed them hotly to no effect. Suddenly seeing me in their house, they snarled and made loud threatening growls in my direction.

"Bad dogs! Shame on bad dogs. Mama doesn't like bad dogs!"

Carol took my hand, led me into a small bedroom, and closed the door behind us. The double bed took up nearly the entire room. I could hear the dogs tearing around the kitchen on the verge of another fight. We kissed briefly, then I sat down on the edge of the bed and we continued fondling each other. Her blouse and bra hung loosely down her front.

While she remained standing, I kissed her belly and tugged her Levi's down over her hips. Shuffling her knees, she let them fall to the floor. She stroked my hair softly and kissed me on top of the head. A low, vicious growl came from under the door.

"Gretch, *shut up!*" Soft whining and the padding of dog feet across the kitchen floor.

"Stupid dogs," Carol whispered.

We continued to kiss and touch and in the touching were carried away, falling in a well of desire, plummeting in its damp wholeness, surrendering to the dark pleasure of pure, uninhibited, animal sex.

Carol stepped out of her clothes, pulling me up to kiss her full on the lips. She slowly unbuttoned my shirt and moved it off my shoulders. Then she reached down, still kissing, and began tussling with my belt buckle. When we finished, my Levi's lay on the floor with hers. Lowering her head slowly, she disappeared into her own hair and began caressing and kissing my belly.

A red dawn of peace and quiet broke on the little cabin.

"Can I get you some breakfast?" Carol asked.

"No, thanks," I said, crawling out of bed and pulling on my pants. "I have to be to work early. How do I get to the Forest Service barracks?"

"You're in luck. It's right down the road, left at the intersection. Then take another left and go a mile and you're there."

"OK," I said, turning to her. "I have to go. It's almost 6:30, and the van leaves for the airport in twenty minutes."

I went to the door, stepping gingerly over the big dogs that lay comatose on the kitchen floor.

"Will you be . . . around awhile?" Carol asked.

I stopped with the door half open and smelled the fresh morning air. Suddenly I wanted to be a long ways away.

"It's hard to say. We never know from one day to the next." I stood there listening to the words and not liking them for how shallow they must have sounded.

"Most likely we'll be here awhile. We'll be staying at the Forest Service barracks or the Hotel McCall."

I looked around. Carol was staring out a window. A small potted geranium bloomed on the sill, the red petals glowing in a band of sunlight.

"I'm usually there by nine in the evening . . . if we don't jump a fire."

I turned to look again.

"See ya," she said.

"Hey, Carol," I said. "Thank you. That was very nice."

With that, I stepped out into a world of pine trees, sky, and meadow grass.

19

The morning after my night with Carol was an unexpectedly quiet one at the base. At ten o'clock I went out to the lawn on the north side of the ready room, rolled out a sleeping bag, and napped in the shade until 11:30. At noon the first load piled into the van and headed for the Si Bueno Mexican restaurant.

Passing Carol's road, I looked down and saw her tiny cottage. What would Sally think now? Writing love letters one night, and hopping in the sack with a complete stranger the next. I hadn't honored the possibility of developing an open and honest relationship with her. Now, it would be different. There would be something I'd always have to keep from her. Just because the major part my life was beyond my control was no excuse to not take responsibility for the part that wasn't.

"Nice you could make it for lunch," Quacks said from the front of the van. "Get a little corn for the duck?"

"I was kidnapped," I said. "It all happened pretty fast."

"Yeah, *right*," Quacks groaned. "Especially after you started kissing her."

"Even blind hogs find an acorn now and then," Troop declared.

After the driver parked the van, the rest of the jumpers unloaded and hit the big front doors of Si Bueno's. Another Forest Service rig pulled in and parked next to ours. Some McCall jumpers got out. One had bandages around his neck and on the side of his face—Frankie Romero.

A month before, Frankie had jumped a nasty, hot two-acre fire in the Nez Perce National Forest with seven others. The fire was burning in an area of tall snags on a rugged, hellishly steep mountainside. Around midnight, Frankie was down near the bottom of the fire, sitting in the dark by himself taking a breather. There came a terrible crash. The rest of the jumpers up the fire line knew that another snag had fallen. A few minutes later a cry for help was heard. Steve Mello, Bob Schumacher, and the rest of the crew hurried down.

By the time Mello, an EMT, reached Frankie, a couple of the other jumpers had cleared away some limbs and debris and wrapped a rag around his neck. Mello confided in me later that he thought the jumpers had put the bandage around Frankie's neck not so much as a treatment but because the wound was so horrible they couldn't stand to look at it. It appeared that he wouldn't live through the night.

Mello had Schumacher hike up the mountain and radio a nearby lookout for help. Next, he removed the rag. A broken branch tip had punctured Frankie's throat right under the cheekbone and ripped a triangle of skin downward to the base of his neck. The V-shaped flap of skin had been forced inside the wound and was tucked under Frankie's collarbone, exposing his windpipe and bleeding neck muscles. About four inches of the jugular vein was exposed, pulsing like a small red snake.

The jumpers hovered at Frankie's side. Out in the wilderness, in the middle of the night, Frankie was going to die. Mello went to work—checking him out, stopping the bleeding. But they couldn't move him to a safe location. The slope was too steep and the terrain too rugged. Instead, they cribbed up a mound under the lower part of his body, built a warming fire, tied his legs together with shirttails and braced his body with rocks. The only first-aid kit on the fire was a belt kit, and it contained absolutely nothing for major trauma. Unless there is a known injury on the jump, only the belt kit is dropped. Some

EMTs carry extra gear themselves, things like bee-sting kits, extra bandages, prescription painkillers.

They treated Frankie Romero for shock, monitored the bleeding, tore up T-shirts for bandages, and took turns holding his head and telling him stories.

"You're going to be OK," Mello told him. "We're not going to let anything happen to you. A chopper's coming first thing in the morning."

Up the hill, more snags fell—one only forty yards away. The jumpers and their fallen comrade were pinned down, unable to move out of harm's way. Mello sent two up the hill to check out all the burning snags that could possibly reach where Frankie lay. If it was too dangerous, they would be forced to move him, regardless.

About two that morning, Frankie's heart rate went from a steady eighty beats per minute to an alarming 115. Mello thought that Frankie was probably dying. He reasoned that the loss of blood had triggered the increase—the heart's final effort to keep Frankie alive. Frankie lay perfectly still, his breathing steady and his heart beating fast. Nothing more could be done. About four o'clock, a Missoula jumper made an offhand comment about the trees in the area of the fire being some of the largest in the Nez Perce National Forest.

"Yeah," Frankie said softly, "and some of the most accurate, too."

"That was such a relief," Mello told me. "I figured, by god, if he can say that and do it with some sense of humor, Frankie will still be here come morning."

It wasn't known then, but the heart-rate increase had been in response to a collapsed lung. At dawn, a helicopter arrived and made a risky pickup off a knife-sharp point while four jumpers hunkered beneath it, holding rocks in place under its skids. Frankie was loaded aboard and flown away to the hospital in Elk City.

After lunch we drove back to the base and found some other jumpers just coming in off of fires. Scott Dewitz, Jim Kelton, and Togie were there, and we talked while they got fire ready. I checked my gear again, then went outside to lie in the shade north of the main building. Rolling a sleeping bag out on the cool grass, I folded one end under for a pillow and lay back and stared at the sky. I needed a nap.

A few other Alaska jumpers came out and arranged bags similarly and began talking about all the lightning that had hit the Payette the night before. Their soothing voices, the sweet-smelling grass, and the coolness of the shade made a comfortable haven from the moment, and I finally drifted off to sleep.

I woke at 2:30 when a retardant plane began cranking its engines. Someone out on the ramp yelled back to the ready room that the plane was rolling to a fire jumped yesterday in Chamberlain Basin. White smoke puffed from its big engines, swirled violently, and then vanished out across the airfield.

Troop opened his arms wide. "Well, hell. Here we are! What're they waiting for?"

Worried that some of our jumpers might be in trouble, those of us on the lawn got to our feet. This was not the time of day, nor of the year for that matter, to be shorthanded on a troublesome fire. From the entryway to the ready room we watched as the DC-7 began its take-off run.

Word came down from dispatch that the jumpers in Chamberlain Basin had lost their fire, that some of their gear had burned. Routine radio traffic, part of which could be overheard at the operations desk, had turned urgent.

"Sounds like the boys up in Chamberlain are getting their butts kicked," Rick Hudson said as he reached for the phone.

"Why don't we go give 'em a hand?" Troop asked. Rick paused.

"I don't know, but they probably think it wouldn't do any good. Maybe they don't want to waste you guys on a fire that's already lost."

"We should go anyway," Troop said. "I'll bet we can catch it."

I stepped over to the deep freeze and lifted the top. "Hey, Troop, you want to be my Nutty Buddy?"

Troop came over, and I grabbed two chocolate drumsticks.

"My treat, Troop," I said, dropping a dollar into the money can. In the next instant the siren let out a hair-raising wail. Hearts pounded as the suit-up area filled with jumpers both suiting up and helping out.

"Hold this a second," I said, handing my Nutty Buddy to a helping jumper. Two minutes later, I was walking heavily out onto the ramp, my jump helmet and reserve in one hand and a gooey Nutty Buddy in the other.

We entered the rear side door and took our places sitting on the floor, single-file between each other's legs, facing the tail. Being number one put me up against the wall opposite the door. John Humphries, our spotter, climbed aboard last.

"Looks like we got trouble," he said as he fastened the safety strap across the open door. The big plane lumbered down the taxiway, stopped at the end, and sat there shuddering while the pilots completed pretakeoff checks. Once the engines had finished their run-ups, they were idled back and the Doug rolled out into the center of the runway. There it stopped, rocking back and forth on its big black tires, all the time Humphries talking nonstop to dispatch on the radio. We sat there for longer than seemed normal, and I began to think that maybe dispatch had called things off. Then the Doug released its brakes and eased into its takeoff roll.

The acceleration tended to slide the jumpers toward the tail. John stood near the door, bracing himself and planting a foot firmly to the floor so that I could place mine against his and act as a holding point for the lineup behind me. As the Doug accelerated down the runway, I watched as the big white and yellow lines moved faster and faster until the Doug rotated, lifted into the air, and left the ground falling away beneath us.

As soon as the engines quieted down, Humphries took up his stance by the door and started talking to the pilots over the headset, alternately glancing out the door at Payette Lake and then back at us.

"We're headed to Chamberlain," he yelled. "Dispatch has lost contact with the jumpers."

John pulled a map out of his spotter's kit, handed it to me, and continued right on talking to the pilots.

The fire was twenty acres and burning hot in the bottom of a large basin when we arrived over it. John was again on the radio, trying to raise someone on the ground. The slope faced west and the afternoon sun had come around full on. The main column had punched through an inversion layer of heavy smoke, and the fire was drawing through it as if through a chimney and making vigorous runs up small side canyons. Only about two hundred yards of perimeter at the bottom of the fire appeared to be cool enough to work with direct attack. Flame lengths on the remainder of the perimeter ran from ten to forty feet in heavy pockets of alpine spruce, ponderosa pine, scattered snags, and thick brush. Occasional flare-ups burst into crown fire all across the head.

The Doug passed near the main smoke column, then slowly came around left, reducing its speed while the jumpers peered out the windows. What I saw was a hell of a lot of hard work in a steep, rugged canyon filled with dry fuels. The time had come. Troop was my jump partner and we were going in.

"George Steele's the fire boss," Humphries yelled to me over the roar of the engines. "Talk to him when you get down there. I'm going to put you guys in on that little flat just down from the fire."

John had talked with George. The crew was safe for the time being; they'd lost some gear. I tightened my leg straps, tucked the map into the top of my PG bag, and snapped the bag on under my reserve. The flat was a good jump spot in one respect—the fire was moving uphill away from it. On the other hand, it was surrounded by tall pines, studded with large boulders, and had a small creek running along one side.

Humphries put two sets of streamers right in the spot, then smiled at me and shouted, "Ready to go to work?"

The Doug pulled around on final.

"There's about three hundred yards of drift," said Humphries. "Most of it's high but bumpy down near the trees. Any questions?"

I shook my head.

"Get in the door."

I completed my four-point check, stepped up, and took a firm grip on the sides of the open door. Troop and I were the guinea pigs. The jumpers farther back in the load would get the benefit of watching us work it out over a tight spot. In the plane, besides Troop and me, we had Quacks, Dewitz, Kelton, Togie, Charlie Brown, Seiler, Firestone, Raudenbush, Dunning, Romanello—all Alaska—plus Rich Nieto and one of his McCall bros I didn't know.

Humphries seemed confident of the lineup. Here we go, I thought. A jump in the Idaho backcountry—tall trees, steep slopes, rocks, big snags, a roaring fire, and my good buddy George Steele waiting on the ground. Now what the hell could be better than that?

"Get ready . . ."

I looked straight off the end of the wingtip and braced myself as tight as I could. Several jumpers in the plane yelled out various ahh-uhhhs, yahoos, and other endorsements. I turned and looked—thumbs-up from all. Humphries's hand slammed into the back of my left leg, and in an instant I launched myself as hard as I could out over Chamberlain Basin.

Jump thousand. From somewhere deep inside the exhilarating roar of an exit, I looked up at the belly and broad wingspan of the Doug and saw Troop clearing the door in what looked like blurry slow motion. *Look thousand.* Right behind him Humphries's head was out, watching to make sure we had good openings. *Reach thousand.* The world spun below me, a blur, fire, smoke, canyons turning. *Wait thousand.* Eyes back to the plane, it disappeared into the sun. *Pull thousand.* I pulled and felt a quick acceleration. As my feet flew up, I was snapped up by my canopy into near-horizontal flight.

I went right to my postopening checks and then oriented to the spot. Once I saw my position and considered what John had said about the wind, I initiated a set of four bomb turns to get vertical separation from Troop. Being the first down, I would be the first to test the wind. If there had been an error in Humphries's judgment, I would be the first to know. I wanted to get low enough to give Troop the room he needed but not so low that I couldn't make the necessary corrections once I got the lower winds figured out. Out of the bomb turns, I looked up and saw Troop five hundred feet above me and above him the second stick of jumpers appearing under fresh canopies.

Below the ridgeline, the wind switched and sucked in more toward the fire, still up-canyon—ideal for a final approach down the draw. At about eight hundred feet I drifted past the spot and saw two yellow-shirted jumpers walking into it. The fire was roaring and popping to my left, its heat uncomfortable through my face mask. In a wide left turn, I came around to quarter down the draw in the direction of the spot, the heat now more at my back.

An up-canyon wind picked up, and I felt the same soaring sensation I'd felt during the practice jump. Still high enough, I effected a full stall and rocked back above the big trees. At about 250 feet, I set up at half brakes—toggles halfway down—and began my gamble with the wind. When entering a small spot in tall timber, you shoot for the very near edge. That way, with an unexpected tailwind, you have the entire spot in which to make adjustments. The risk, of course, is that if a head wind suddenly comes up and blows you back, you may not reach the spot at all.

The giant pines began to define the structure of the airspace above the spot. A particularly large one stood directly in my path, so I eased left, cleared the edge of the timber and moved out over a gaping hole in the forest. I was still awfully high for the room I had. Across the spot another one hundred feet was a wall of trees. I reefed down on my toggles, going into deeper brakes but not wanting to risk a stall. I floated for a moment, suspended, then settled forward easily, saw the ground racing up, went to full brakes, flared the canopy across a small patch of dead grass, and felt my feet hit the ground.

With slightly bent knees, I twisted my upper body to the right so that my left hip and thigh took the impact, then forced my momentum over onto my butt and back, somersaulted over my left shoulder and came up on my feet. I threw off my helmet and let out a great yeehaw.

"All right," the two jumpers on the ground cried out in unison. I looked up at George Steele and Kenny Franz, BLM jumpers out of Boise.

"Dammit," I said. "There's nothing like doing something right when you have to."

Not everyone had managed to land in the clearing, but all had made it down safely. Our cargo came in, and while the rest of the crew gathered it up, George filled me in on what had happened.

"We got ten guys here, jumped yesterday and cut line all night. It was pretty brushy, so by this morning we hadn't made it all the way around. All night the fire kept just ahead of us, making little runs on the south side of the ridge in snowbrush, big logs, and shit. We hung in, hoping we might have a chance to corral it up on the ridge, but then about four hours ago she punched through the inversion and took off. We lost all our line except the very bottom."

George stopped talking, jacked his jaw a couple times, then started laughing. His face was lined with soot and dirt and his eyes were bloodshot.

"As we headed back to the jump spot to get our gear, it torched up near the bottom, and a big log rolled across the line and started a bunch of spot fires below us. Within minutes, they'd burned together and made two runs up the hill, flanking the area where our gear was. Our cargo was dropped close to the fire so we could hit it quick, but it turned out to be too close. We saved our jump gear and all the chutes, but most of the cargo burned up. We got a chain saw, some food and water, but that's it. We backed off and regrouped, and now they're building line up the left flank. You guys take the right. We might still catch it if it doesn't slop over the top too far."

George paused, took off his glasses, spit on them, pulled out his handkerchief, and began cleaning the lenses. Some older jumpers still call George "Slasher"—a nickname earned when he had accidentally stumbled into an ash pit and badly burned his calf. George had whipped out his knife, cut away the burned flesh, smeared chain-saw oil on the wound and bandaged it tightly with the lower half of his T-shirt.

But that was a minor injury compared to his accident in 1985. On final into a jump spot at ten thousand feet, in the rugged mountains of northern Nevada, George's chute oscillated radically and shot him full speed into a rock the size of a pickup truck.

"When I regained consciousness I didn't feel all that bad," he once told me. "Then I felt something warm in my mouth and spit out a couple teeth. My jaw didn't come together right, so I knew it was broken. I tried to stand up, but a pain shot through my right leg, so I figured it was broken and shifted my weight to my left leg and it *was* broken, so I put a hand out to steady myself on a rock and damned if my wrist wasn't broken, too."

Before his jump partner could get to him, George got on the radio to the spotter in the plane: "I got a broken jaw . . . a broken leg . . . a broken wrist . . . and I lost a few teeth . . . and now I think I'm going to pass out."

Which he did.

George spent that winter in a rehabilitation program that included sixty-six hours of dental work and reconstructive surgery to his jaw. By the following spring he was well enough to pass the PT test, running three miles in under twenty-two and a half minutes and earning himself a slot back on the Boise jump list.

"It'll be an all-nighter, but what the hell," George grinned. "We haven't got anything better to do."

We agreed. The saw teams would press ahead, falling the most hazardous snags, clearing brush, and bucking the downed logs out of the way so that a fire line could be built by those that followed. Our reasoning was that if things went well, we would have our initial containment lines in by morning and perhaps corner the fire on the rocky ridge. If our lines could be widened enough before the heat of the next day, if the winds didn't get too radical, *and* if the spots over the ridge could be picked up in time, then we could probably hold the fire at around eighty acres.

We knew we were talking a lot of ifs, and that further, *if* things didn't go our way, then the fire would go big and require a fire management team, major amounts of taxpayer dollars, and the mobilization of hundreds.

As George and I finished discussing our plan, Charlie Brown asked if I'd like to team up with him falling snags. We packed up a Stihl chain saw fitted with a thirty-inch bar, a complete saw pack including extra gas and oil, and stuffed our PG bags with cans of beans, corn, beef jerky, instant coffee, and candy bars.

George's bunch had secured most of the bottom perimeter by digging a fire line five feet wide down to mineral soil. By dark we had secured the rest of the tail and started moving up. I scouted ahead, selecting the most dangerous snags, those that would likely fall during the night. Charlie ran the saw first, cutting out the big logs and brush. He was an expert saw man, having had experience as "lead saw" during his days with the Chena Hotshots. We moved upslope, clearing the heaviest stuff, while the line builders came behind.

The forest was primarily mature alpine spruce, with some stands hit hard by spruce budworm. In areas as large as a football field, every tree was dead and had been for several years. Tinder-dry, they burned hot clear to the top, leaving huge slabs of glowing bark and sapwood barely clinging to their sides. Massive burning limbs hung tenuously, ready to break off and come crashing down.

Snags kill more wildland firefighters than any other single hazard, and these were big, hot, and ready to go. Some were five feet thick at the butt. Using Pulaskis, Charlie and I would scrape away the burning debris from around the bottom to cool the base area enough to get in close. Then I'd stand back eight or ten feet acting as lookout as Charlie buried the chain-saw bar into the tree. When anything broke loose, I'd yell and he'd drop the saw and jump back out of the way.

Charlie's handsome dark face was wet with sweat. Firelight shimmered on his cheeks and sparkled in his eyes.

"Murrmanski," Charlie said, looking directly into the heart of the burning forest. "This is what I love."

A forest burning at night is a beautiful sight. Fire on that scale creates scenes that few people will ever see or even imagine.

"Fighting fire makes me feel alive. You know what I mean, Murrman?"

A moment passed while we both stared into the interior of the fire.

"Yes," I said, glancing at Charlie. "I know what you mean."

A headlamp flashed down through the smoke. A voice hooted in the night. We hooted back.

"Hey, great," George said, coming closer. "You guys are making good progress."

"How much more to the top?" Charlie asked.

"Oh, maybe a quarter-mile slope distance, but the fire edge is so crooked. It might be more."

George looked down the hill at the rest of our crew digging line.

"I'm glad you're this far up. The guys on the other side are almost to the top. There's at least a dozen spot fires over the ridge, but most of them will burn out in the rocks. We got plenty of work, though. That's for damn sure."

"I'm going to swing around the bottom and then check those guys on the other side again," George said. "I'll see you on top sometime in the morning."

George went off down the mountain. Charlie cranked up the saw and started bucking a log. A flare-up close to the line hissed and roared and stung our faces. I could smell the ends of my hair starting to burn. We ducked to the ground behind some brush to wait it out. Later, another searing blast drove us back into the cool darkness of the forest again, where we caught our breath and watched from the shadows. Down the hill a hundred yards, we saw the headlamps of our line builders flashing this way and that in smoke and dust. They were progressing slowly, but progressing just the same.

By eleven o'clock, the crown fire near the head had subsided to an intermittent running ground fire. Our section of line had cooled into a forest of glowing coals, with small flames licking yellow, orange, and red. Occasionally in the interior of the fire, one tree would fall into another and burst into flame, sending clouds of sparks rolling up through black shafts of trees to soar hundreds of feet into an oily black sky.

We were about a half mile from the clearing where we had jumped, but the crooked fire line was nearly a mile long. The left-flank crew had tied in to the rocks. They were spreading out back along their line to improve it and work the spot fires.

I started falling snags at midnight. Just after one we decided to break for a bite to eat. Charlie lugged the saw pack plus the two PG bags up the fire line. The higher we climbed, the steeper it got.

"Here's the place, Murrman," a joyful Charlie bellowed. "I don't know about you, Murrmanski, but I'm hungry."

I set the saw down and looked up the hill at him. Charlie Brown—his father a black man, his mother a Mexican woman—stood in the light of the dying fire like a bronze statue. Broad shoulders, narrow waist—boldness, vitality, and grit.

"Murr?" Charlie asked, full of himself, smiling broadly. "I'm a hungry man. Do you how hungry I am, Murr?"

Charlie's unrestrained love for life, for our fire on the mountain, and at that moment, I think, even for me, was almost awkward.

"OK," I conceded. "I give up. Just how fucking hungry are you, Char?"

"I'm so hungry, Murrmanski, that I could eat that chain saw you're packin'. I could just pour a little oil over it and eat the whole thing. Maybe wash it down with a can of gas. You know what I mean?"

I carried the saw up to where he was waiting and threw it at his feet.

"Well, feel free," I said. "I'm tired of running this bitch anyway."

"When the goin' gets tough, Murrman . . . ," Charlie scolded gently.

"I know," I said. "When the goin' gets tough, the tough head for camp."

"We got miles yet, Murr. This is no time to puss out."

"You think we'll be OK here?" I asked, looking around for snags.

"Piece of cake, Murr!" Charlie sang. "Piece . . . of . . . cake!"

My saw partner then pulled up a log that had rolled out of the fire, chipped some coals off one end, tossed on some kindling, and we had an instant cooking fire. Before long, two cans of pork and beans were bubbling over. I took out a package of beef jerky and a Snickers bar.

We ate the beans. They tasted like filet mignon. When they were gone, I wiped out my can with a handful of moss, filled it with water, opened three packets of freeze-dried coffee, put two in the can, and dumped the other in my mouth and washed it down with a swallow from my canteen. Placing the can in the hot coals, I leaned back against a rock and waited. Once I drank the can of coffee, I immediately started feeling better.

"We better get back to work," I said. "I'm starting to get chilled."

"Stoved up, Murr," Charlie groaned, getting to his feet. "Stoved up is more like it."

It was late—1:45. Soon we'd begin to feel sleepy. First, the unpleasant, dull aching in the knees, then the mild nausea. Staying active would help. Either way, sleepy, aching, or whatever, it was time to press on.

We continued up the fire's edge with me dropping snags and Charlie swamping. Everything was going well until about 2:30, when I noticed a new area of stars appearing overhead. The smoke column was switching from the northeast to the south—a night wind began to blow. Smoke rolled down out of the fire and across the line. The interior of the burn rekindled hot. My radio came alive.

"Seiler, Firestone."

"This is Seiler. Go."

"This changes things," Jack said. "We're gonna have to pull back and widen the line."

Charlie and I packed up our gear and headed down. When we met the rest of the right-flank jumpers they were gathered on a small flat where a spur ridge dropped off to the south, to our left—away from the main fire.

"We've got a snag burning down in this draw," Seiler told us. "It's dropping sparks all over the place. Eventually its going to fall, and if the fire gets hot enough it's going to make a run flanking the main fire. Jack's bunch have dropped back toward the tail. It could cut us off from them *and* our escape route."

A big crash sounded in the dark below, and a huge ball of fire rose above the trees.

"There she goes," Al said. "I'll take three guys—Romo, Bush, and Troop. We'll go down and see what it looks like. The rest of you stay here and hold what we've got."

Al and his three dropped off the spur ridge south, building line down as they went. Charlie started the saw to cut a wider path so the diggers could widen the main line, and I pulled debris and threw it out of the way. In heavy smoke, our eyes burned, stinging with sweat, gritty with ash. An hour passed as we worked, choking and coughing and occasionally even laughing.

"Firestone, Seiler," we heard on the radio.

"Go ahead, Al."

"Can you send us a little help? We've got line down to this slopover, but it's trying to flank around into the next draw over. We've got to hold it to this one or it'll get into a big patch of bug trees and take off."

"Copy that," Jack said. "But I'm not comfortable leaving the tail, and the guys up the left flank are committed on top. We're scattered too thin."

Mitch radioed Seiler, offering to head down with Togie and Dewitz.

"Sounds good," Seiler said. "We've got to hold this draw if we can. It's only about eighty yards of line, but if we lose it, we'll lose this whole side."

Mitch took Togie and Dewitz and headed down. Charlie and I stayed and patrolled the line, pushing hot stuff back into the fire and watching for spots. The fire line along the right flank came up the hill, then crossed the edge of the little flat, and that was as far as it had been dug. It had been intended to angle on up the hill toward where Charlie and I had eaten, but now with the slopover in the draw that would have to change. The little flat extended out from the main fire and formed the head of the draw where Al and his group were working. An orange glow was steadily building in the bottom of the draw. Two chain saws howled frantically somewhere near its center.

"Charlie," I said. "Let's build a second line along the edge of this flat. That way we can burn out the top of the draw. Even if they hold the bottom, she's going to make a run up through here. If it comes hard enough, the heat from the main fire could combine with it and roll it back out over the woods, dropping sparks back into that draw Al's trying to keep it out of."

"Murrmanski. You sly old dog. Now you're thinking. Once we get a line across this flat we'll carry it down a ways. That'll help contain our burn out, plus give Al's bunch something to tie in to as they come up."

"Another thing," I said. "We need to save this flat if we can. It might turn out to be our only safety zone."

Charlie and I began digging line down the spur ridge away from the fire, with only our headlamps for light. Soon we were in a rhythm, dripping with sweat, muscles taut and aching, breathing ash and dust and smoke, not talking, just working. Another hour passed. We'd made about sixty yards and tied in to a rocky outcropping.

The light in the bottom near Al's bunch suddenly flared bright, and a loud hissing sound rose up from below. While Charlie headed back to the main fire line, I walked down another twenty yards through a surreal world of sillouetted black pillars, limbs, branches, and smaller trees—all about to be consumed by fire. Tiny wisps of smoke appeared everywhere, rising up from the forest floor. Spot fires, not much bigger than a human hand, waved their frantic warnings in the unfriendly darkness as I passed by. Atop a small cliff I held up and took a moment to witness the final moments of a three-hundred-year-old stand of alpine spruce.

From where I stood, it was about a hundred yards to the bottom of the draw where Al's spot fire burned in a flat three times the size of a volleyball court. Al's bunch had looped around the bottom of the fire and were digging line uphill toward me. The upper end of the fire was making small upslope runs, building heat, gaining momentum. Tree branches whipped back and forth in superheated air. Sparks rained upward, becoming lodged in branches. Moss-covered trees burst into eerie blue flames. Fire spread rapidly into the crowns. Suddenly three large spruce torched off and flames whipped through their crowns in a great roar. To be that close to something that powerful, to gaze down into the tops of giants, to see the shimmering heat as the surrounding forest appeared out of the darkness lit by an ominous orange cast sent a wave of immense exhilaration through me. Showers of embers blew upslope, starting more spot fires at the base of the cliff. Smoke swirled crazily around me. I turned and hurried back up the line Charlie and I had cut, through the shadowy coolness and up the spur ridge. When I got there, I found Charlie widening the main fire line.

"We've got to start the burnout now!" I yelled, digging into my PG bag for a fusee. "She's starting to make up. A couple more runs and this whole draw's gonna go."

Charlie started on one side of the flat and I on the other. We burned along below the little flat until we got to where the line dropped off down to the rock outcropping. Then we began contour strip-burning, taking about ten feet each strip, so as not to cause too much heat at one time. In ten minutes we had the entire top burned out, halfway down to the rock outcropping.

"This is either going to work like a dream or we're going to get the shit scared out of us," I remarked offhandedly to Charlie.

"Murrman, we'd better get our butts up top in case all this goes to shit, and we have to move out."

By the time we got back to the little flat, a menacing roar and a great orange glow rose up out of the draw.

"If she spots toward the ridge, our way out is going to have to be back down the right flank," I yelled. I hated saying it since it clearly meant that we might be forced to try to find somewhere inside the main burn in which to hole up.

Charlie and I quickly hustled down the main fire line thirty yards, then turned to wait. Suddenly the fire thundered up out of the draw, crowning in the big timber. The lower end of our burnout came alive and stood up in a crown fire of its own. The ground began to tremble. Somewhere in my head, common sense screamed the obvious. All our effort was going to be pitifully inadequate. For one small moment Charlie and I turned to look at each other, stunned. The world around us was on fire. Spot fires ignited in the small flat. Then, in a spectacular turn of events, the fire racing up the draw began pulling our burnout fire down toward it. A huge rush of flames slammed together just uphill from the rock outcropping and sent clouds of sparks flying three hundred feet into the sky. Smoke was pulled from all directions. Charlie let out a wild howl and moved up into the small flat and began stomping out small spot fires. I rushed to the far side where our new hand line dropped down the hill and did the same. For the next half hour we ran down spot fires and reinforced our lines. When we finally met up again, we high-fived each other, then fell into a laughing fit. After we calmed down, I pulled out my watch. It was 3:45.

By 4:30 the fire in the draw had died down, leaving the little canyon no longer a threat to the right flank. Al and his bunch had arrived at the bottom of the rock outcropping, having dug two hundred yards of hot line around the bottom and up the hill to tie in to ours. Charlie and I remained in the area of the little flat, checking our line down to the outcropping, improving it, and patrolling back and forth for spots.

"Fucking amazing!" Al yelled as his bunch crested the flat. We lined up and high-fived one another until everyone had high-fived everyone else. Bush, Romanello, Mitch, Dunning, Al, Togie, Troop, Dewitz, Charlie, and I stood together, sweat-soaked, shoulder to shoulder, marveling at the fire corralled in the little canyon.

"It was incredible," Charlie said, laughing. "That was some great shit."

"Man, oh man," Al said. "When she went out the top, I thought, oh, well, there she goes."

Firestone and the rest of the jumpers were on the radio. Everybody had had a tough run of it, but we'd outlasted the wind and outmaneuvered a fire that had been ready to blow out the right flank. After a few more quick reviews of what had happened, the others spread out along our lines once more to patrol and improve them.

At 5:30, by the first light of dawn, Charlie and I watched the last snag break off the stump with a loud pop. After it crashed, there was silence.

In a small clearing near the ridgetop Charlie found a cluster of granite boulders that looked like a group of sleeping elephants. In between two we located a sandy area with a panoramic view of the fire. All but the last stars had disappeared. Charlie made a campfire while I cleared a space to lie down.

I got out my rain tarp and sweatshirt, then propped up my PG bag for a pillow. I lay down on my tarp a few feet from Charlie's fire, pulled the hood of my sweatshirt over my head, and got comfortable. Charlie sat cross-legged, staring into his fire. Flickering light played across his face. Smoke and sparks curled skyward.

August 29

Stiff and cold, I woke just as the sun cleared the ridge. A thin layer of smoke drifted over the canyon where we'd jumped. Inside the fire area a few stumps, stobs, and big logs were still burning. All around us were mountains, their broad slopes spiked and tufted with trees. To the south, the soft light of dawn played upon granite peaks.

I gathered up some sticks and relit the fire. Charlie woke up, blinked his eyes, and grinned.

"Murry, my friend," he said warmly. "This is what I love."

"Pretty marvelous, isn't it?"

"No damn doubt about it," Charlie said. "We're living a dream."

Walking back down the fire line, Charlie and I came to the combat zone of trampled grass, cut logs, and brush piles that marked the division between the blackened and standing green forest. The fire had calmed. A good fire line was in place. Loaded heavily with our gear, we worked our way down to the small flat of the burnout. The right-flank guys were hunkered around a campfire, laughing and sipping cups of coffee, and talking about a snag that had fallen in the early morning.

"Well, you know," Gary Dunning said, "it was around six or so. Everybody was thinking about getting a little sleep. Jack and I couldn't sleep, so we were checking the line. It's a good thing it cracked like that when it did, 'cause we needed a head start. Otherwise, I tell ya, man— you would have still been digging."

Dunning glanced at a massive snag lying straight down the fire line, its foot-thick branches rammed into the ground.

"We knew it was a snag, but when we turned to look we couldn't tell which one. Then we saw it. Of course, it had to be the biggest, nastiest piece of shit around," Dunning said, wry resignation in his voice.

The butt of the big spruce had shattered when it hit, exposing slick bare areas of sapwood. The heart was rotten. Hidden inside what otherwise appeared to be a healthy tree, the heart had decayed and fallen in on itself, creating a cavity. Sometimes you see the telltale conks, the fruiting structures of the fungi that indicate rot. Other times, the conks have withered away or been burned off. Charlie and I had examined the snag and judged it to be sound. Our job had been to cut down the hazardous snags, so you can imagine what a relief it was that no one was hurt.

"It wasn't real pretty," Firestone added, partly amused. "It was coming right for us. We started yelling to warn everybody, then everybody started yelling. We beat feet down the fire line. I knew in my mind about when it should hit, but I didn't know if it was tall enough to get us. It was close, though. We ran a third of its length, and the top hit fifteen feet behind us."

After a bite to eat, Charlie and I hiked up along the ridge and ran into quite a few new firefighters. Helicopters were delivering crews. We found a few more troublesome snags and dumped them. By two that afternoon, the word was out. The Chamberlain Basin fire was contained and the jumpers were being released. By four we were all back in the clearing where we had jumped twenty-four hours before. Quacks who had hung up on his jump in, was high in big ponderosa, cussing and sawing away, trying to retrieve his chute. After all the cargo chutes and gear was collected, we kicked back in the shade on our gear bags, napping, eating, laughing, and waiting for our ride.

August 30

Back on flat ground, I could tell that my gimped-up knee had gotten worse. It had swollen a little and it hurt, but at least it felt strong. My other knee was beginning to hurt as well. Afraid someone might notice me limping and put me off the jump list, I concentrated on walking and standing as normally as I could. The knee seemed all right in action, but once things slowed down, the inflammation and pain inevitably returned. It was time again for the smokejumper's faithful ally—ibuprofen. I went to the EMT room and got a fistful. Once I had my jump gear ready, I stopped by operations to check on the weather.

"Damn good job, you guys," John Humphries said. The jumpers placing their gear on the speed racks chuckled to themselves, bemused.

"No, I'm serious," John insisted. "We got a call direct from the fire via upstairs dispatch. They said it was an outstanding effort and the new fire boss wanted you all to know that he appreciated it."

Smokejumpers tend to make light of any praise that comes from outside the experience, but inwardly we value it a great deal.

"Hell," Humphries said, "you may as well take it. Think of all the other times when we did the same thing, but no one was there but us."

I checked the bulletin board and found a note that Carol had called. There was a cardboard box on the floor marked ALASKA MAIL. I rummaged through and I found a letter from Sally.

The letter from Sally was exactly what I was hoping for. She was having a good time with her summer in McCarthy, had one more month to work, and was considering traveling with me in the fall. Once we spent time traveling together, I felt certain that she would at least accept a ride back to the Lower 48. We could travel the Alcan Highway down through Banff and Jasper. Then on to my place in Northern California to check it out. If all that happened, I was sure that once she saw my beautiful log home, the forty acres, and the love-liness of Quartz Valley, no way would she be apt to go anywhere for a while. Spending a winter with Sally could be a turning point in both our lives. We could take a midwinter trip to Mexico and travel the west coast, camp on the beaches in my camper, swim, get suntans.

All I had to do was keep away from Carol. There was no point in feeling guilty. It had just been a one-night fling, and I would keep it that way.

20

September 4   Moose Creek, Idaho

I dropped my pack in the shaded dry grass under a big Douglas fir and waited for my jump partner to come into view on the trail below. In the headwaters of Moose Creek it was dead calm and hot as blazes on the south-facing slopes. Across the canyon the higher ridges and mountains were still in bright sunlight while the lower country disappeared in blue afternoon shadows. At the same elevation, downstream six miles, I could see Goat Peak and the area of our fire. Tom Romanello and I had jumped it five days before, just at dark as we were flying back to McCall from a dry run some Zulies—Missoula jumpers—had beat us to. Our flight home had been a routine plan B— we were going home. I was lying in the back of the plane near the door.

Without warning, the Twin Otter had lowered a wing into a tight turn, and I opened my eyes to see the other jumpers looking out the windows. I twisted my body around on the floor and stuck my head out the door. Straight as a pencil, a thin column of smoke rose up out of a deep, narrow canyon. My heart started pounding as I looked up and saw the spotter coming back to the rear of the plane. Back to plan A—we were going to a fire.

"Nice little two-manner," the spotter yelled. "We'll have to hurry."

We were running out of daylight. I couldn't see the bottom of the canyon, but upslope a ways there was a small clearing. He threw one set of streamers. I never did see them after they were halfway down, and I don't believe he did, either. We circled once more then pulled up on final.

"Take that little clearing," the spotter yelled, pointing into the blue void. "There's no wind. We'll fly you a line to the fire after we drop your cargo. You'll be about four hundred yards uphill from it."

He took hold of the toe of my left boot and moved it out over the threshold of the door. Lying facedown, he peered forward under the plane and spoke into his microphone. The ship crabbed left. He moved my foot farther out over the edge.

"Watch for rocks under the brush. Get ready!"

I got the slap and stepped into the wind with Tom right behind me. Once the chute opened I got right to work steering toward a small green opening in a sea of blue trees. That close to dark is no time to get injured because there would be no way to get out before morning. Halfway down, Tom and I passed from sunlight into shadows, and I could see the ground more clearly. Ten seconds later, I began to notice the tallness of the trees, their lofty crowns defining great high spaces above the ground. The relative positions of the treetops shifted below me in an ever-changing pattern, telling me moment to moment that no matter what, it would be terrifying to miss the spot and sink into them. The spot was about sixty by sixty feet and at least a hundred feet deep. A tall alpine spruce sat partway out in it. The lay of the land on the hillside became clearer. A buckskin snag leaning into the spot came into view—a terrible white ghost with long brittle branches and a spear-pointed top. I looked over at Romo. He was behind and above me about fifty yards. At three hundred feet I spotted big gray rocks hidden in the brush. Turning left, I moved back toward bare ground. Holding right, I tried to not crowd the space Romo would be shooting for. Rocks appeared in what had first looked like open ground. Weaving left, weaving right, I oscillated wildly, aiming for a place without boulders. The top of the tall alpine spruce snagged the right side of my canopy. I swung out and around, the top snapped, and I began to fall, then was jerked backward violently into a green blur. During a barrage of cracking sounds, I tumbled back out to the edge of the branches, fell a few more feet, and finally hung up solid. Just then, Romo sailed past right into the middle of the clearing.

Moments later, while I was still cussing and thrashing around eighty feet up preparing for my letdown, Romo yelled. "Hold it. Hold still. Wait'll I get your picture."

I flipped him the finger. "Show this to friends," I growled. My chute was punctured clear through, so I knew it wouldn't pull out. Taking the end of my letdown rope, I ran it through the D rings on my pants, and tied it off to my tight riser with three half hitches. I snugged up the slack, took the rest of the rope out of my pocket and dropped it to the ground, then made my final check for lines around my neck. Slowly I pulled my cutaway clutch, releasing me from my risers, and fell a couple feet, bouncing up and down Howdy Doody–like. Romanello laughed some more. Down the rope I descended, brushing by the wider branches nearer the bottom, until I swung into clear air and reached the ground.

"Well, at least there wasn't any cow shit up there," Romo chuckled. He tried taking another picture, so I threw the rest of my letdown rope at him.

For the next four days we worked the Goat Peak fire, digging in the soft fragrant soil, "potato-patching" the entire burn. About the size of a baseball diamond, the fire was creeping in the forest duff. Some ash pits were deep and very hot. The entire area had to be turned with shovel and Pulaski four times. Without a handy water source, we kept digging until the dirt finally cooled enough to extinguish the hot coals.

The fourth day a helicopter dropped a cargo net in our jump spot and returned later to pick it up. We sent as much out as we could, since we would have to pack the rest. Jumpers usually pack out of wildernesses, since helicopter flight is restricted within wilderness boundaries.

When jumpers are in great demand, they sometimes make exceptions and fly us out.

We kept our sleeping bags, a shovel, a Pulaski, and all the food we had—which wasn't much. Jumpers jump with three days' food, and we'd eaten nearly all of it. Ray Brown, Rick Blanton, and four others had jumped a fire near the bottom of the canyon. Ray radioed to tell us they were moving out and would cache some food in a small spring along the trail, a gesture we appreciated.

The next morning we got up at dawn, packed, made one final check on the fire, and headed for the bottom of Moose Creek. Just below the jump spot we found a game trail that led to an old Forest Service trail. Four miles from our fire we came upon some red flagging marking the spring and the food cache: two mushy apples, a water-logged orange, and three cans of beer. Not bad if you were looking for a quick buzz and something to feed the pigs.

Sitting there waiting on the switchbacks for Romo, I took out the map again. As near as I could tell we'd covered fourteen miles, coming down the ridge into Battle Creek and then downstream to its junction with Moose Creek. We had stopped at the creek to eat, drink the beers, take a swim, and marvel at steelhead two feet long. It was 1:30 when we started up the Moose Creek trail, and I knew then that it was going to be tough making Elk Summit by dark.

Romo came up the trail, swaying along, his T-shirt drenched with sweat. He stopped to catch his breath and stared down at me like a wild man.

"You're a tough son of a bitch," he said, exhaling an agonized grunt as he slammed his pack to the ground. Pulling off his T-shirt, he hung it on a small pine tree, then flopped down on his pack, swiped his forehead with the tips of his fingers, and flung the moisture to the ground.

"How much food you got?" Romo asked.

I took out a cornflake bar the size of a deck of cards, unwrapped it, and broke it in half.

"This is it," I said. We'd shared a small can of beans that morning before starting out. Since then, just the beer, the orange, and the apples.

"Take these," I said, handing over two small white pills. "Ibuprofen and pork and beans, buddy. That's all we got."

"Yeah," he said, grinning. "Minus the pork and beans."

I checked my watch—5:15 P.M. We shouldered our packs and started out again. Elk Summit lay somewhere up and beyond the rimrock that jutted above us. Every step was uphill and steep. Thanks to the helicopter, our packs weighed only forty-five pounds compared to the normal one hundred. We were tired and hot but weren't really suffering the miserable hell of a full pack out—the cutting shoulder straps, the back pain, the neck pain, the knee pain, the not-even-a-mule-would-be-this-damn-dumb pain.

As daylight faded, our hunger grew and our exhaustion with it. More trail and more trail, energy reserves gone. Our bodies weren't burning food anymore, not even fat. They were burning muscle.

At 7:30 we topped out at a junction with another trail and ran into a pack string of several horses and a half-dozen mules. They were resting. The packer was a leathery old fossil with a bent back, a bald head, and a crimped cowboy hat.

"You bet," he said, wiping his head with his handkerchief while casting a loving eye in the direction of his shiny black mules. "I packed many a jumper outta this country, by Jesus. We're overloaded now or I'd put your stuff on one of these old girls."

Two chubby, pink-skinned men in their mid-twenties sat on two of the horses. I thought about asking them for food but decided I'd rather not.

The old packer chuckled and said, "I always told the Forest Service that if they'd just take a little time and train my mules to parachute, they'd have an improved smokejumper."

"I tell you," Romo said, looking up at him, "about now a mule could sure as hell have my job."

We asked if they had any extra water. They did, and we had long drinks from the old man's canteen.

"Ain't you a might old for this smokejumping?" the packer asked.

"Way too old," I said. "But I haven't got enough sense to quit."

"Probably ain't got any quit in you. Most of you jumpers don't. That's the way with me, too, though. I'd rather do this packing than anything I can think of . . . even though it sure as hell gets rough."

We thanked him for the water and told him that we had to keep moving.

"Well then," the old fellow said, tipping the brim of his hat to us, "we'll leave you boys to your misery. You about got 'er made, though. Two, three more miles is all."

Off they rode, leaving us following behind. The sun set low in the west. Yellow dust hung over the trail, and the canyon began filling with the same blue haze we'd jumped into five days before.

The surface of Hoodoo Lake was as smooth as black marble as we came up on it and began flanking its west shore. Here and there, water circles bloomed as fish brushed the surface. The sweet smell of the shoreline was a welcome change. Up ahead we could see a parking lot, the unloaded pack string, some pickups, and horse trailers. When we arrived and dropped our packs for the last time in our fourteen-hour, seventeen-mile pack out, it was the moment we had longed for all day. We looked each other over, broke into smiles, and shook hands.

"How's your knee?" Romo asked.

"It's fucked," I said. "That last two miles made mush of it."

The old packer approached us with four blue ice-cold cans of Hamm's beer. Our arms shot out like rattlesnakes. I drank the first one in two swallows. The second one took a bit longer, but as soon as it was gone, the packer lifted his chin toward the back of his pickup.

By 9:00 the packer had all his stock loaded, and at 9:15 a man arrived in a Forest Service pickup. "You must be the jumpers out from Goat Peak," he said.

We threw our gear bags in the back, thanked the packer, and climbed into the front seat. The road from Hoodoo Lake was an old hunting road, crooked and studded with large boulders—it was low gear all the way, weaving around rocks under the big, dark timber in the moonlight. We made small talk with the driver for a while, then Romanello and I collapsed against each other, sound asleep.

September 5

The clock above the operations desk read 6:45 P.M. Fifteen more minutes and I would have the night off, a clean bed, and a full night's sleep. Jumpers milled around, passing back and forth through the big doors of the ready room as they did every day at that time.

A few minutes before seven I tried Carol again, deciding to give it up if she didn't answer. I wasn't sure what I would say anyway. She had called twice while I was at Moose Creek. Maybe we could have dinner and I could explain that I was involved with someone else. That would be the right thing to do. There was something essentially wrong about supposedly being in love with one woman and participating in a cross-country love affair with another. On the other hand, the memory of Carol standing in front of me with her large, full, firm breasts was one I had been reviewing a lot. She'd called. I had to call her back.

Carol answered.

"Hello, Carol?"

Carol was happy that I'd called. She asked if I'd been on a fire.

"Yes!" I said. "In Moose Creek."

Carol wanted to know where Moose Creek was, so I traced our trip out backward, the best I could clear back to Goat Peak. There wasn't much else to talk about. Most of what we knew about each other was hardly phone material.

"You must be tired," she said. "You sound kind of funny."

"Well, yeah, we're all pretty tired."

"Could we meet at the yacht club for a drink?" she asked.

"Yes, I'd like that very much."

We said good-bye and hung up. I found myself standing there staring out the bay doors of the ready room, thinking of what I had just said.

Snails have demonstrated greater backbone.

September 6

Halyards snap against the masts of the small sailboats moored in the marina as a light breeze blows in off Payette Lake. The western sky blends from yellow to blue as the moon appears above the east ridge. Pleasant clatter and chatter spill out the back door of the restaurant as it works its way through the dinner hour.

I sit at a table waiting for Carol and massaging my left knee. Back from Moose Creek, we'd found Troop hopping around on crutches after bruising his hip on a jump in the Bitterroots. That afternoon Chip Houde had cannonballed a mountain near Whitebird and was being medevacked to the hospital in McCall for X rays. During a letdown, a Missoula jumper had mistakenly tied off to his harness and fallen forty-five feet and suffered a compound fracture of the left tibia.

Taking a sip of beer, I turned and saw Carol stepping out onto the deck with a big smile on her face.

"Hi, stranger," she said.

"Hello. Have a seat." I got up and pulled a chair out.

"I thought maybe you'd run off with Smokey Bear."

Carol was prettier than I remembered, with dark auburn hair that complimented her creamy light skin. She seemed happy.

"Carol. Thanks for calling and leaving the message. It's nice to see you."

The waitress came and Carol ordered a margarita. We talked about her week and her life and a little about Moose Creek. There was something awkward, yet erotic, about sitting there getting to know her after the way we'd been together that first night. Carol's hair curled down along the sides of her face and lay out across her shoulders evenly.

"How are the dogs?"

"Oh," she said, embarrassed. "They're the same . . . a lot of trouble. Night before last, they jumped the neighbor's fence, treed their cats, and knocked over a rabbit hutch. The dogcatcher has them now."

I almost laughed out loud. "Oh, that's too bad," I said.

The waitress arrived with a margarita big enough that she had to carry it with both hands.

"I don't have the money to bail them out. I barely make rent now. Poor things, they haven't any place to run. They can't stay in the house . . . it's, well, you saw the place."

She stopped for a second as if attempting to catch her breath.

"Anyway, I keep them tied up in summer and inside some in the winter. Now and then, I let them run for a few minutes, and sometimes they don't come back. Mom says I should get rid of them, or put them back out on the ranch. But then I'd miss them. I guess I need a bigger place, one with a fenced-in yard."

I was sorry that I'd brought up the subject of Carol's dogs.

"How long do you think you'll be here?" she asked, as she reached down with both hands and lifted the margarita carefully to her mouth.

"I don't know. Two weeks, maybe three. Depends on the fires."

"Will you come back?"

"Oh sure! I come here every winter and spend time skiing with friends."

She sipped at the salted rim of the glass, set it back on the table, then looked out over the lake.

"I told Mom about you."

"You did?" I asked, almost gasping.

"She says to never get involved with a smokejumper. She says they move around too much, and they're always leaving for someplace you've never heard of and will never tell you when they're coming back, and there's not one that can tell the difference between love and a fence post."

Suddenly I wished we were talking about the dogs again.

"Well," I said. "I'm sorry your mother feels that way."

"It's OK. It's just because she had a boyfriend once, and he was a smokejumper. His name was Catfish."

Catfish, I thought. *Oh my god.*

"Well," I said again. "Excuse me. I'll be back in a second."

When I came back from the rest room Carol was gone.

I went outside to look around and ran head-on into a group of jumpers pressing through the front door of the yacht club.

"Hey, old man. What happened to your date?" Charlie Brown asked.

"The one that got away," Seiler said ruefully. "Another one."

Not wasting any more time on the subject of my love life, they moved on in the direction of the bar, leaving me standing on the sidewalk.

"You better enjoy this place while you can," Charlie said over his shoulder. "They took a pounding over in Montana this afternoon. Rumor has it some of us may be heading for Missoula first thing in the morning."

"Oh, well," I mumbled. "There's lightning in Montana."

The idea of having another beer was appealing, but I needed a break from smokejumpers. I felt bad about the way Carol had vanished. I walked up Main Street until I got beyond the town lights, then turned and walked back on the other side. Enough time had elapsed that I thought I could call her place and see if she was there. The phone just kept ringing. Not only had I betrayed my own sense of loyalty to Sally by deciding to bed up with Carol again, but now that I had revealed what a spineless worm I was, I was going home alone anyway. I decided to go back to the hotel and forget the whole thing. I was tired, I was limping, and I was acting like a fool. If we were going to Missoula, the opportunity for a full night's sleep shouldn't be passed up.

I turned the key, and the brass doorknob felt soothing in my hand as I entered my room. I groaned as I looked upon the big queen-size bed. I had been fairly sure of spending another night with Carol. Sure enough that I had gone ahead and paid, out of my own pocket, the difference between the cost of sharing a room with another jumper and having one to myself. What a big bed. Lovely pillows and such nice lace. In the corner of the room were my work clothes, boots, and travel bag—all of it beat-up and grubby. I looked in the mirror and saw myself, stringy, muscular, tanned, and turning gray. I undressed, keeping an eye on the man in the mirror. *Catfish*, I thought to myself. *That son of a gun. We sure used to have a lot of fun together.*

Turning the lamp down, I threw back the comforter and sheets and crawled in. The night before, Romanello and I had slept on the lawn of the National Guard Armory in Grangeville. Three nights before, we'd hunkered in the dark woods of Moose Creek. Where would I be by this time tomorrow night? I shut my eyes and tried to sleep but couldn't.

Thoughts of Montana flashed in my mind. I imagined the lightning at that very moment starting new fires, fires that would determine the course of the following day and where I would next be bedding down. Unable to sleep, I turned the light back on and started a letter to Sally.

After I finished the letter and clicked out the light, there came a soft rapping at the door. I wrapped myself in a towel and stumbled over to answer it. *Oh god,* I thought, *not Montana—not at this hour.*

"Hi," she said. "Sorry. I know it's late."

"Carol," I said, standing back and offering her a space in the room. "Are you OK?"

She wasn't. As soon as I shut the door behind her, she began crying.

"Wait, I'll get the light," I said. I tucked my towel tightly around my waist and made my way over to the nightstand. With a click, soft light flooded the bed, leaving the rest of the room in gray shadows. Carol remained standing by the door, wiping her eyes with a small handkerchief.

"I'm so embarrassed," she said.

"Carol. It's all right. Really. Are you OK?"

"I don't know."

I went over to her, put my arms around her, and she started crying harder.

After a minute she stopped crying. She reached out, hugging me back, alternately stopping to wipe her nose.

"It's all right. It's not that late," I said, pushing some of her hair back off her forehead.

"I thought I'd better leave while I could. I don't even know you, and still I missed you while you were gone, and now you're going to leave again." She stopped, blew her nose, then sniffed loudly.

"I thought I'd better take mother's advice and forget the whole thing. That's when I decided to leave. Sorry. I could have said good-bye."

How on earth, I thought, could anything so weird be happening to me? A woman I hardly knew, whose mother hated me and my friends, was crying in my hotel room in the middle of fire season. Even more bizarre, somehow it felt right that it was all my fault. In a few minutes Carol dried her eyes and began talking easily.

"So, I went for a walk," she said, sniffing. "I didn't want to go home. I just wanted to be alone. I walked down to Lardo's, then back to town."

She stopped to gather herself.

"I went back by the yacht club and saw some of the other ones on your crew. They said you were walking out of town when they last saw you. That's when they told me you'll probably be going to Missoula in the morning." She stopped talking and we hugged some more.

"That's when I knew I wanted to see you, even if it was just for tonight."

We began kissing. I backed up and sat down on the bed, leaving her standing in front of me, looking down, her brown eyes streaked with eye shadow.

As the last light went out that night at the old McCall Hotel, we lay wrapped in each other's arms, strangers to each other in every way.

September 9

Life had settled down in McCall. No major bust activity, just the steady work that jumpers get with late-season lightning. No one was sent to Missoula, at least not from McCall. Grangeville sent eight and Boise ten. I was glad not to go. I'd been there several times before, and it was nice to have a break in the action and some time with Carol. I phoned the Payette County Animal Shelter and sprung her two hairy ones, charging their bail and board to my credit card, then phoned her and told her that they were ready to be picked up. She was so happy she started crying.

The days were getting cooler, with warm afternoons and brisk nights. The routine at the base was just right for getting the work done and still having time to enjoy a private moment with a book or writing in my journal. Word had come from the jumpers in Redmond that Tyler Robinson had sideswiped a ponderosa pine, fallen out of it, and hurt his back. Tyler had been taken to the hospital in Bend. Chip Houde was back with us after spending a night in the hospital. The doctor had examined his knee and placed him on light duty. Troop had traded in his crutches for a spot back on the suit-up racks. The crew was generally rested.

Today started off with a relaxed roll call and then PT, including twenty minutes of stretching my leg muscles on the lawn. Buck Nelson, Dewitz, Seiler, Baumgartner, Charlie Brown, and Togie wanted me to take the afternoon off and go with them to rent a speedboat for waterskiing on Payette Lake. We were engaged in that discussion when the call came in—two loads to French Creek.

While we suited up, Rick Hudson stood up behind the operations desk and gave us the lowdown. "This fire's already got two guys on it," Rick yelled as the aircraft outside on the ramp began running up their engines. "They jumped it yesterday but lost it a little while ago. It's over in the Salmon River breaks, near French Creek. Fire moves fast down there, so be heads-up."

By the time the Doug leveled off at three thousand feet over the north end of Payette Lake, we could already see the smoke.

"This fire's west of French Creek" the spotter yelled. "About a mile from its junction with the river."

We were going to an ugly fire—a dangerous one—one to catch the first night or fail and watch it go big.

In that part of the country, the Salmon River—the River of No Return—flows east to west in a gorge contained by two seven-thousand-foot ridge systems. Culminating in stands of old-growth Douglas fir, tamarack, pine, and spruce, the highlands are relatively level but gradually steepen into tawny, broad-shouldered ridges of grass, brush, and scattered timber, to plunge steeper and steeper into narrow draws, eventually falling six thousand feet into the bluffs, cliffs, and rockslides of the canyon bottom itself. French Creek joins the river from the south. Scattered along the river, a half mile below the fire, were several summer homes.

When we arrived, the fire lay on the south side of the river just below midslope, extending uphill a quarter of a mile. It had burned about fifteen acres.

Jeff Bass looked at me, then quickly turned his eyes back out the door.

"Welcome to smokejumper hell," he yelled.

Jeff had been a McCall jumper for several years before coming to Alaska. We would rely heavily on the McCall guys and their experience with fire in the breaks. Besides the two McCall jumpers, John Humphries and Gay Johnson, we had Troop, Jeff Bass, Firestone, Quacks, Mitch, Seiler, Romanello, Charlie Brown, Jim Kelton, and me—all Alaskans.

We circled north of the fire, out over the canyon, and waited while the Twin Otter moved in for a closer look. The Otter was having trouble finding flying room between the smoke column and one of the main ridges that paralleled the river to the south. Dropping us on the lee side of the ridge was ruled out for fear of downdrafts. The fire looked to be about two thousand feet above the elevation of the river. The slopes below it were steep and broken—not safe to jump. Above the fire, in heavy timber, the slopes weren't as steep, but the risk of jumping directly uphill from a fire is generally unacceptable. The Doug came around upslope between the ridge and the fire and fell in behind the Twin Otter. Everyone got to the left-side windows.

As the Doug slipped into the shadow of the smoke column, it suddenly shuddered and slammed down heavily. I knelt near the door and hung on. I could look directly down on the fire. The smell of woodsmoke filled the plane. The spotter stood in the door and talked nonstop to the pilots. There were a few spot fires near the head. The bottom of the fire was a jagged creeping line that fingered up and down the hill with occasional small flare-ups. The west or right flank (looking uphill from the bottom) carried flames from four to six feet high and was eating its way across the contour to the west. Each finger would occasionally make an aggressive uphill run and then, because of the sparseness of the fuels—grass and scrub brush—slow down and rekindle enough heat to run again.

I looked at our group. All the lightheartedness of our day back at the base was gone. In its place had come an anxious, contemplative dread, fueled by the character of the fire, the absence of a jump spot, the slam of the turbulence, and the familiar smell of woodsmoke.

Clearing the shadow of the smoke column, we came around into the sun on the north side. The decision was made to fly a crosswind pattern east to west, flying directly above the river, and drop jumpers in a small spot three hundred yards off the left flank, east of the top of the fire. The fire wasn't directly downslope, but the fuels were dry and winds in the breaks can change at any time. The problem was we had little to pick from. It was not standing-on-its-head steep, but it was far from level. The size of a baseball diamond, granite boulders lay scattered in the upper half. Tall timber formed a wall around it. Immediately south of the jump spot, the mountain climbed steeply to the ridge, above which now drifted a massive roll of copper-colored smoke. Below the spot, the mountain fell away down yellow spur ridges separated by brushy ravines.

The Otter dropped streamers first. They drifted south up the slope, long strands of yellow, blue, and red descending gracefully over the canyon, until suddenly, in an accelerated arc, they sucked in toward the fire, tumbled in spirals, blew sideways, and disappeared in the smoke.

The Doug and the Otter flew in a circle, working off each other, flying at opposite points on a great compass. Now it was our turn. Our streamers did like the others. Each spotter made two more streamer runs, trying to adjust for the drift. Quacks got sick, partly missed his plastic sack, and puked on himself. The smell filled the plane. The assistant spotter grabbed a rag from behind the tail bulkhead, wiped off the floor and an embarrassed Quacks then tossed the rag out the door. On the last attempt, the streamers hit fairly close to the small clearing, and our destiny was sealed.

The spotter, kneeling next to the door, held up two fingers and yelled, "Let's do it."

Humphries and Johnson would be jumping first, Troop and I next, then Bass and Firestone. The first two would be jumping rounds from fifteen hundred feet; squares from three thousand. The main smoke column swept upslope, then rose vertically. At a thousand feet above the jump spot, it curled back north and drifted lazily over the fire.

"Be careful," the spotter shouted at Humphries. "You've got three hundred yards of drift . . . just like the smoke. Stay upwind if you can. Any questions?"

Humphries and Johnson had none and were soon hooked up and taking their positions in the door. I looked at Troop. "You ready, buddy?"

Troop already had his helmet on. Inside his face mask all I could see was whiskers and eyes.

"Let's go get her," whiskers and eyes shouted.

We pulled on final, hit some turbulence, and were jolted to the right. The plane corrected left.

"Get ready, John." Humphries braced in the door.

Around came the spotter's hand against the back of Humphries's left calf, and suddenly the two jumpers were gone. I finished pulling on my gloves, stood up, tightened my leg straps, and hooked up. As we climbed to three thousand feet I watched the first jumpers from the Twin Otter disappear in the smoke.

Our spotter repeated the wind and jump-spot information. Humphries and Johnson were down by the time we came around, so I couldn't tell if they'd made it, either. The best way in appeared to be from the southeast, quartering north and west. The tricky part would be to not misjudge the lower elevation winds and end up short, long, or wherever, in the trees.

"Any questions?" the spotter asked.

We had none. It was straightforward. Play it tight—make the spot.

"OK. Get in the door," he yelled, patting the threshold with his hand.

I hooked up my drogue static line, completed my four-point check, and took my place in the door, with the toe of my left boot extending three inches over the edge. *Get ready to fly,* I told myself. I could feel Troop's hands on the back of my main as he steadied himself behind me. The Doug leveled off on final, flying upriver. I sat looking straight out at the sidehill as the fire came into view under the wing. Since we would be jumping out at fifty-five hundred feet above the canyon bottom, and the spot was up the slope twenty-five hundred feet, the perception was of jumping into an enormous space and drifting sideways onto a high shelf.

"Get ready," the spotter yelled.

When I got the slap, I lunged out into the slipstream and was sucked up by it and blown away. Falling, I looked back to see Troop coming out the door and the shape of the big plane moving away.

As soon as I got a full canopy, I started my safety checks and began steering. Down below, the jumpers from the Otter drifted upwind of the spot, to the left of the fire. The first stick from the Otter had missed the spot and hung up in a small stand of fifty-foot Douglas firs bordering the side away from the fire. Two canopies lay in the spot— Humphries's and Johnson's.

In four quick bomb turns I established vertical separation from Troop. By the end of the second revolution, the canopy was standing on its side and the centrifugal force had my body almost parallel to the horizon. Coming out of the last turn, I made sure I flared facing away from the fire.

I settled into the shadow of the smoke column. Various twists and bends of the Salmon River turned red-orange and yellow through the smoke. The fire roared and swooshed; the river rushed among rocks and spilled over waterfalls, the sounds blending in the air.

I swung around and caught a glance of Troop hanging high behind me. He had his camera out as his parachute flew unguided straight for the fire. At about six hundred feet the wind, doing just what the streamers had indicated, drew strongly upslope. In the turbulence my chute began to pitch sideways and rock forward and back. I lined up on the spot and planned to enter it low from the northeast, just clearing the big pines on the eastern edge.

The area nearest the jump spot had several spots crowning simultaneously. The roar grew louder. Flames fifty feet high whipped back and forth. I thought about Troop, who was higher and closer in to the fire. I eased down on my right toggle and came around slowly over the big timber, then began my descent, aiming for the near edge of the spot. The tops of the trees loomed below only a hundred feet, then fifty. Then I cleared the timber and entered a gaping hole with the ground still a hundred feet down. Instantly I was below the treetops, and the wind quit. The chute smoothed out, moved forward, and down I came, flaring easily and landing right on top of Humphries's canopy. Ripping off my helmet, I looked up for my jump partner. Troop was on final at three hundred feet, hands on his toggles, dropping steadily. Pulling my cutaway clutch, I released the risers to my main and spit on the ground, trying to rid my mouth of the metallic taste of adrenaline.

The rest of the jumpers came down in the little clearing or some-where nearby, and soon the area was littered with parachutes. Ro-manello ran long and slammed into the base of a large pine and was knocked unconscious. Seiler checked him out. "Once those bluebirds quit circling, he'll be all right," Seiler told Humphries.

Quacks had trouble again. He snagged the side of another big pine, was jerked sideways into it, and crashed down about fifteen feet, then caught and ended up mad as hell swinging back and forth forty feet off the ground. The Doug and the Otter zoomed overhead dropping cargo. We gathered the cargo and jump gear, stacked it in the center of the meadow, cleared an area ten feet wide down to mineral soil around it and then made our plan.

Initially a two-manner, the fire had been restricted to an old pon-derosa snag and its immediate surroundings, but the hillside was steep, and burning material had kept rolling down. Then the afternoon winds came up. Losing their fire, the two original jumpers had radioed dispatch for help, but now something was wrong with their radio, and we had trouble making contact. All we could hear was the breaking of their squelch when they tried to answer. Humphries suggested they an-swer yes with one click, no with two. By that means John determined that they were at the bottom of the fire and would meet us there. I looked at my watch. It was 3:30. We still had four to five hours of day-time burning conditions ahead of us.

Humphries would be in charge of the west side, and George Steele, out of the Twin Otter load, would handle the east. On hillsides, fires usually run uphill, leaving the tail at the bottom and the head up top. The flanks are established from the tail point of view looking uphill into the fire. In this case, the right flank was the west side, the left flank was the east. By the time we got the cargo gathered and the clearing cut around our gear pile, the fire had burned within fifty yards of the jump spot. It was decided that Steele and the Twin Otter jumpers would drop down the east side to a point near the bottom, where they could tie in with the original jumpers and begin working back up. Our load would stay and protect the meadow, then work the west side. Standard procedure is to begin at the tail—in this case, the bottom—then work up the flanks. Since our fire was so close to the jump spot, half of us had to stay and protect our gear.

"We got retardant coming," Humphries told us.

In the meantime we widened the clearing around the gear pile and began digging line along the edge of the jump spot nearest the fire. Smoke rolled high, darkening the sky and causing an uneasy calm in the clearing. Ash began to fall silently.

We all knew that we were compromising some safety issues, but our only other option was to abandon our gear. At the time, the fire behavior wasn't so extreme that we couldn't use the jump spot as a safety zone. We extended our fire line west, contouring across the slope above the fire to halt the advance of the head, and then north downhill between the fire and the area immediately below the jump spot to keep it from rushing up at the spot. Just before the retardant arrived, we began burning out in the vicinity of our gear pile. About sixty yards down the hill, a whirlwind materialized inside the burn. It flew around in circles blowing ash and dust, then sucked itself up into a twenty-foot vortex of fire and started toward us. As if alive and planning what to do next, it moved up to a big ponderosa pine, tucked itself in on the uphill side, and stopped to build bigger. Fanning the area into a fury of sparks and flying coals, it began picking up sticks and pinecones, shooting them into the air, and then moved away from the big pine and proceeded up the hill, bringing the fire with it.

"This is bad medicine," Bass said.

The fire devil made three side trips back and forth below us before moving into the main fire. Just then the retardant ship showed up and made a direct hit on the hottest part of the fire immediately below the jump spot. Johnson, Seiler, Troop, and Brown rushed down and began scratching line around the area where the fire devil had spread fire.

"That was a warning, boys," I heard Bass say.

Ten minutes later another retardant ship arrived and knocked the head of the fire down.

"Let's leave Romanello here to watch the gear," Humphries said. "The rest of us will work across the head and take advantage of the retardant. We'll build line and burn out as we go."

The old pine snag stood lifeless and grotesque against the smoky afternoon sky, its limbs contorted and twisted back on themselves. From its trunk, slabs of gray sapwood had fallen away, exposing the rust-colored, pitch-impregnated heart. A lofty monarch of three hundred years, it had died a summer death long ago, killed by fire. Sap moved up from the roots had lost its lift, stalled, and had been drawn back down into the tree's heartwood, where it eventually solidified into tons of pure pitch. The debris of its own destruction lay scattered about its base—sapwood slabs, limbs, bark, its broken top. Staghorn lichen bristled yellow-green up the reach of the old snag's charcoal-plated north side, where a century of ant colonies had made their home. Woodpecker holes riddled its midsection. Fifty feet up the hill, the tail of the French Creek fire crept steadily through scattered clumps of cheatgrass and scrub sage.

We dug hand line two hundred yards across the head, staying well inside the retardant and burning out where necessary. Mitch did most of the burning, using fusees from the fire packs to light the edge of our line once it was finished. By burning out the fuel next to the line while the conditions were right, we increased the holding capacity of the line later on, if the conditions worsened. We rounded the head and extended our fire line along the upper west flank in an attempt to keep the fire from crossing. That worked for the first fifty yards or so, but the fire kept flanking west, and we were eating a lot of smoke. We had secured our gear for the moment, but as far as corralling the fire was concerned, we had accomplished little, since a large portion of the lower perimeter was unmanned. I looked at my watch. It was 6:20.

Nearing what should have been the end of the burning period, the fire continued to burn hot.

"This is stupid," Bass said matter-of-factly. "What're we doing here?"

"Let's head back up," Humphries called up the hill.

Back at the jump spot we all pretty much agreed that as long as the current burning conditions remained as they were, it would be crazy to continue our line downhill. We decided to grab a bite to eat, to wait, and to watch.

A helicopter passed over the jump spot and flew out over the canyon to recon the fire. After directing Steele and his jumpers, the fire officer aboard suggested that our group return to the west side and continue digging line and burning out. Our line was holding, he said. John told him that we had tried that and that it had seemed too risky. John pointed out that the unsecured line in the bottom left it possible for the fire to run uphill at us, but the man said that the fuels were sparse, and that it was only about four hundred yards to tie in with the other jumpers. The winds had died down, he said. John asked him for a size-up of the topography below our line, and he confirmed what we already knew—it was steep and it had fire on it. The helicopter circled a few more times, addressing Steele's group, then flew away upriver.

The old snag leaned out over its rocky perch. Below it, heavy brush covered a slope that pitched off into a slide area of loose rock, then ran a hundred yards into the bottom of a V-shaped draw. The creeping fire had reached the bottom of the old snag, burning into the deep stockpile of duff and cast-off debris. Flames began licking their way up the trunk and into the pitchy heart. Moments later, the snag was a rolling boil of tarry, black smoke.

We packed up again and started down the hill. Humphries, Johnson, Charlie Brown, Troop, and Seiler returned to where they had left off digging line, while Bass and I came behind, widening it. Jim Kelton was next behind us, with Mitch, Quacks, and Firestone behind him completing the burnout. Romanello's ankle had begun to swell, so he would stay on top to protect the gear and patrol the adjacent line.

We fell into our line-digging rhythm as John scouted ahead. The wind at the time was light, flowing out of the canyon parallel to the fire's edge. The timber was scattered and thinning as we descended. The sun dropped below the horizon. Our hillside of fire calmed and cooled as the sky burned red. I took out my watch. It was 8:30.

We kept on digging and the country kept getting steeper. In some places it was necessary to dig footholds with our Pulaskis. Quacks came down the line with a five-gallon cubitainer of water. When he got to Mitch, he slipped and dropped it. It began rolling, and by the time it reached Bass and me it was going too fast to stop.

"Look out!" Bass yelled. The group down the hill scattered as the rogue cubie went flying by them, disappearing over the brow of the incline.

"Dirty, rotten, son of a bitch," screamed Quacks. "All the way up this steep-ass bastard to the *fucking* jump spot, eating *fucking* smoke all the way, and carrying it all the way back down and then dropping the fucker as soon as I get here."

Cussing and ranting at himself, Quacks went off up the fire line again to get another cubie.

The old snag continued to burn in a fury of its own making. The top had burned off and had fallen into the brush below. Fire shot out of woodpecker holes in every direction. The lean of what remained of its sixty-foot trunk increased slightly, causing its top to tip a quick three feet to the west. At ground level the fire had burned halfway through its base.

By 9:30 we'd made it about four hundred yards down the west flank. The burned-out areas were to act as our safety zones. It was near dark. A few stars came out. Bass turned to me.

"Did you feel that?" he asked. "We just got a wind switch."

The smoke was slowly turning, drifting downslope.

"The minute it changed I felt it on the back of my neck," Bass said. "Nice and cool."

Sweating like we were, the cooler high-country air certainly felt good, and there was a sense that this was indeed the change we'd been waiting for—the diurnal wind switch from up-canyon to down-canyon. I took my headlamp and strapped it on my hard hat. Working on, we steadily dropped down, now and then hearing George's group on the radio discussing the amount of rolling stuff that kept crossing their line. The west flank kept burning in strips adjacent to our line and making occasional thirty- to forty-foot runs. Most of the interior of the fire had cooled pretty well, although islands of unburned fuel remained farther in.

By building line downhill with fire below, we were breaking one of the most critical rules of wildland fire fighting. But, in the distance we had come since leaving the area of the last retardant drop, the angle of the line had been steadily improving in our favor—we were able to keep the fire more to our right, all the time taking the line straighter down the hill. In another hour or so we would hook the bottom and tie in with the others.

Then the wind quit altogether.

Bass and I stopped and watched the main fire. Down the hill the headlamp beams of our leaders flashed back and forth in the smoke, then froze. Up the hill, too, headlamps shined weakly out into the darkness, not moving. I turned to Jeff.

"Something's not right," I said. "What's the deal with this calm?"

"You never know in the breaks," he answered.

The old snag had become a standing column of red coals, spitting occasional sparks up into the black night. The cavity at its base emitted a pulsing glow, orange and pink; blue flames exited bright yellow. The ancient pillar stood its last moment, a stark silhouette against the burning mountain. A groan issued forth from somewhere in the earth, and the snag began to fall. Slowly at first, then faster, with sparks streaming from its top; then downward off the rocky point to rip free of its root crown and slam to ground, burst into flames anew, and slide downhill, stringing fire into the V-shaped draw.

"Sounded like a snag," I yelled to Humphries. He'd heard it too and was on the radio to the guys in the bottom.

"I don't like this," Bass said, loud enough that Humphries could hear.

"Yeah. I don't either," came the immediate reply.

Then we spotted it. A glow in the black below. We couldn't see fire, just a growing light and an orange smoke column getting brighter. Fire sounds began swooshing and hissing up the hill. Then, all of a sudden, a great roar rose out of the darkness and surfaced in a tongue of blue-orange fire.

"Which way is our safety zone?" Quacks yelled.

Bass looked up the hill. "The main fire's gonna reburn, so you can forget that," he said, his voice charged with authority. "Our last good zone was about a hundred yards up the hill, but we can't get there fast enough. Move out away from the fire. When we're far enough out, we'll drop down around the bottom."

In a span of a couple of minutes, events slipped out of gear. An up-slope wind began to blow as quickly as if a switch had been flipped. There was a confusing radio transmission with numerous voices yelling in the background. Yellow flames broke the brow of the slope and shot thirty feet into the air. The next thing I knew I was trotting across the rough hillside with Jeff Bass right below me and Gay Johnson right behind Jeff. Jim Kelton was behind me. People were yelling both above and below. The smoke hit us broadside, thick and hot, and thousands of sparks blew furiously all around.

"I'm going up," Kelton yelled.

"Not up! This way," Jeff yelled back to him. "Up's too steep."

Kelton disappeared into the smoke. The country in front of us was a series of small draws, each a hundred feet or so wide with brush, grass, mule-ears, and very few trees. When we reached the far side of the first draw, we looked back. The entire bottom half of the fire and all the perimeter below where we had been working was now a wall of flames fifty feet high. Islands of unburned fuels inside the main fire erupted into pockets of crown fire, roaring, and rekindling the areas already burned.

The country lit up so that I could make out individual trees three miles across the canyon. The smoke was on us again, frightening, hot, and burning our throats. I thought about Romanello, his swollen ankle, and the jump spot and our gear and our puny little fire line scratched down to mineral soil.

Large firebrands—sticks, branches, pinecones—began raining down all around us. We broke into a run, stumbling on rocks and brush, thrashing ahead. When we cleared the crest of the second draw, terror jolted through me. A ragged finger of fire was ripping up the far slope of the third draw, nearly as high as we were. The draw below had become a raging inferno. I heard Romanello yelling over the radio.

"Hump. It's right below you. Get the hell outta there."

"We're moving," Humphries said, out of breath. "I think we can clear the ridge in a couple minutes."

"If you're not outta there in forty-five seconds," Romo came back, "you're not going to make it."

"Don't fall down," Bass yelled over his shoulder. "Don't fall down."

I don't think any of us had yet panicked. It was all too fast. I wasn't about to panic myself, but I did sense that across that same hillside, stumbling in the smoke, becoming confused, some of the others might be. I slipped and slid in the loose shale, every few steps dropping my uphill hand to keep from going down. Through the brush we charged, dead limbs snagging our shirts, PG bags, and pant legs. We ran a pace of negotiated agony—too slow and you burn to death, too fast and you fall and burn to death. Training images of fire entrapments shot through my mind. I remembered how some of those fleeing had made it to within twenty feet of safety, only to be suddenly knocked to their knees by superheated air. Flailing, their lungs seared, pack straps melting on their backs, they stumbled their last steps to that place where they vanished forever into the world of fire.

My awareness narrowed to the immediate few feet in front of me. Only minutes before I'd been aware of the canyon, the sky, the fire, and everyone on the west flank and their safety and well-being. The stakes had quickly run so high that I could no longer spare concern for anyone's safety but my own. A horrible recognition. The last stage before hysteria.

Smoke and sparks. The heat was burning my bare wrists and neck. I could smell singed hair. I began to feel confused. I was running with Jeff Bass and another guy, dodging rocks and fire somewhere in Idaho. A great light raised high in the sky behind us, a roar at our backs, and the space in front of me instantly filled with flying embers and extreme heat.

"Oh, shit," I heard myself blurt out. Our escape route was on fire. We had lost contact with everyone. We were running in fire. I had often imagined what it must be like to be trapped in fire. The blinding heat, the horror. The fire is on you. Your body is burning. That detached observer within recognizes that your worst fear has, in fact, become your final reality. You are finished. You are burning to death.

"Don't fall down, goddamn it," Jeff yelled angrily. He reached around and messed with his web gear and PG bag, but I couldn't tell what he was doing. Maybe he was thinking of going for his fire shelter.

Then he and Gay were running again, picking their way through a gauntlet of spot fires; I ran right behind them. When we got to the finger of fire in the third draw, we kept on the contour and moved right on through, arms held up before our faces as five-foot flames singed our hair. Forty, fifty feet to the other side, we broke out into the night air, retching from the smoke.

We stepped up the pace. There were still fifty yards of spot fires we had to weave through. Below, to our great relief, we could see fire shirts—Troop, Humphries, Brown, and Seiler. We ran on beyond the third draw and dropped down to join them. Jim Kelton quartered down across the sidehill to join us, his eyes wild, barely able to speak.

"Where's Jack and Mitch and Quacks?" Bass asked, his chest heaving.

Jim shook his head.

"I . . . don't know," he said. "They tried . . . going up."

Too out of breath to talk, we fell in line and started a steep, angling descent away from the great light, out into the darkness of the canyon, and on until we were well below the main run of the west flank. I don't think ten minutes had passed since we'd left the line. Farther down the mountain we pulled up on a small bench that overlooked the bottom of the canyon. There we took a breath and drank water. We were stunned. Humphries hauled out his radio.

"Steele, this is Humphries." There was no answer.

I looked at Bass. He stood looking up the hill at the fire, his face drenched in sweat. He glanced at me.

"This ain't no fuckin' good," he said.

I looked up the hill. Mitch, Quacks, Firestone, and Romo must be up there somewhere.

"Firestone, Humphries," John called.

"Jeff," I said. "Thanks for keeping your head."

He shot a look at me as if I were crazy, then shook his head. "I didn't want you guys to fall. That's all I could think of. Damn it! There wasn't time. If any one of us had gone down, there wasn't time to stop. You know what I mean? I hate to say it, but there wasn't time."

"Anybody on the French Creek fire, this is Humphries. Does anybody copy?"

"Humphries, Romanello."

"Tom, this is John. We made it. We're going to drop down into the bottom. What's your location?"

"I'm in the jump spot with the gear."

"Have you seen anyone?" Humphries asked. "The fire's gone to shit, people are scattered all over. I need a head count."

"Negative. I haven't seen or heard anybody."

"Are you going to be all right?"

Romanello had held tight in the jump spot. With the earlier burnouts, the retardant, and the relative flatness of the jump spot, the fire had burned around him.

"We lost some chutes in the trees, but I'll be OK as long as a snag doesn't get me. There's a few burn holes in these gear bags, though."

"All right. Thanks for keeping an eye on us," Humphries said. "Let me know if you hear or see anybody."

John tried his radio several times, but no voices came back. All we could hear was the sound of a big fire roaring up the hill.

Charlie Brown took out his handkerchief and began wiping blood off his neck. I went over to him.

"The dogs on the saw," he said. "I didn't realize it until we stopped."

During our run, Charlie had dropped his saw but not before it ripped into the side of his neck. I pulled the collar of his fire shirt aside and saw three bloody puncture wounds about an inch apart on the back of his neck. I tore a strip off the bottom of my T-shirt and using Charlie's handkerchief tied a bandage around his neck.

"Let's head to the river," Humphries said. "We can't do any good up here."

A Forest Service lookout on some nearby mountain called and said he could see the light of our fire and asked if we needed help relaying messages. It was a tough situation for Humphries. He would have to tell McCall that we were missing some people—that he hadn't been able to account for everyone. That much he was obligated to do, but neither did he want to cause unnecessary alarm.

It took an hour to find our way down through the shadowy draws and ravines into the river bottom. Bass slipped in some shale and banged his leg. We held up, and Humphries came over to check it out. Jeff's knee had taken a direct impact off the sharp point of a rock.

"It's not that bad," he chuckled, faking a smile. "But it sure hurts like hell."

Charlie Brown's bandage was red with blood and had begun dripping down the front of his shirt. During our trip to the river, John tried several times to raise someone on the radio, but no one answered.

In the bottom of the canyon we hit a single-lane dirt road and walked upriver until we were passing directly beneath the fire. Once again on flat ground, I could tell how much I favored my hurt knee. Since the fire call I hadn't paid much attention to it, but now the old grinding pain was back. From the top of my PG bag, I took a small vial of ibuprofen and swallowed six.

The bottom of the canyon was two hundred yards wide, and the river took up about half. The fire and its great pall of smoke flooded the entire area in an unearthly orange glow. Out in the river, white water crested pink while canyon walls loomed, transfixed in the wild and unholy light.

We headed up the road. In a quarter of a mile we came to a house. No one was there, so we took a break, ate a bite, and filled our canteens from a faucet in the front yard. The flat where the road ran was timbered with old-growth ponderosa pines, some of them five feet in diameter, their cathedral columns magnificent in the shadowy light. Humphries told us that he was going to scout up the road. I pulled out my watch. It was 12:10.

Kelton was looking at Bass, trying to get his attention. The group was outwardly calm but inwardly filled with dread. Jim wanted relief. We were safe, but what about the others? Why weren't they answering their radios?

"This is pretty bad," Jim said to Bass.

Bass started in talking the instant he heard Kelton's voice.

"When it made that last run," Bass said, "the last I saw, you were heading uphill, and all I could think was ... he's a dead man. Jim's a friend and a good guy, but he's a dead man."

"I didn't make it very far before I could see you were right," Kelton said. "Mitch and those other guys went up, too, but I lost them in the smoke." Kelton hesitated and turned something over in his mind, then said, "They must have made it to the jump spot by now."

Bass flashed a look at Kelton, then me, but didn't say anything.

I tore more off my T-shirt and made another bandage for Charlie's neck. After we had finished eating, we dug a fire line around the back of the house. Then we walked up the road another quarter of a mile and ran into Humphries.

John was somewhat relieved. He had made radio contact with George and the other east-side jumpers. They were holing up on a rock bluff about halfway down the east flank, safe for the time being. Steele guessed the fire to be at least five hundred acres. None of their line had held. John had also run into the two jumpers that had originally jumped the fire—they were up the road digging line around a second house. Everyone had been accounted for except Mitch, Quacks, and Firestone.

Humphries called us together.

"Here's the deal," he explained, clearly still shaken. "I talked to George's bunch. One of their guys has a sprained ankle. They're sending someone down with him. Mad Dog and Olson are down, too, just up the road. I called McCall. Engines are on their way around through Riggins. They should be here in a couple hours. I talked to Romanello again. Things have cooled down around the jump spot. He took a quick hike down our line to see if he could find Mitch and those guys. He only got a little ways and didn't find anything. We're going up to start a search."

"Charlie," Humphries continued, "you stay here and patrol these houses. Build some more line if you can. Get with the engine guys when they come. The rest will go up."

Humphries hesitated for a moment, thinking. Just as he started to speak, we heard something. Between the sounds of the river and the distant roar of the fire, there came another sound—a recurrent crashing thud.

"Rock," Seiler yelled. "It's a rock."

Looking up into the shadows and straining to hear, each of us tried to determine the direction the rock was coming from. We all figured it out at the same time—straight at us. The group scattered. I ran behind a big ponderosa along with Bass and Charlie Brown. We lined up behind each other and peeked out. For a moment there was more yelling, then it was all listening. Awful crashing noises were coming right for us. A giant *kathump, kathump* shook the ground and the big tree.

Then I saw it. In one great arcing flight, a boulder the size of a Volkswagen came hurtling through the air, sheared the top out of a large pine, hit the road with an enormous earth-pounding boom, bounced up through the dark, and cannonballed into the Salmon River. Smaller rocks came streaking down, landing everywhere, rushing by us and hitting some of the trees. The main rock had passed not forty feet from where we hunkered behind the big ponderosa.

Once all the rolling and crashing and splashing had stopped, we walked to the road and stood at the edge of an impact crater four feet deep and eight feet across.

"Wow. That was great," Troop said happily. "Did you see that baby fly?"

"See it?" Bass growled. "We're lucky the son of a bitch didn't kill us all. This ain't like watching a movie, Troop."

"We're going to have to fill it in some," Seiler said. "Those engines are going to need to get by here."

"Well, if they do, they better park behind a tree," Troop beamed.

"You guys go ahead," Charlie Brown said. "I'll fix the road."

Humphries gathered us up again.

"We've got to find those guys," John said. "We can't waste any more time. Get ready to do some hiking."

Of the twenty on the fire, Romo was protecting the gear, one of Steele's group had a twisted ankle, Bass had a banged-up leg, and Charlie Brown a bleeding neck. Mad Dog and Olson, the two original jumpers, would stay on the road and work with the engines, protecting the summer homes. Three were missing. That left eleven possible searchers. Bass, of course, wouldn't hear of remaining behind, so our search group was twelve.

We made our way out of the bottom, moving up through a sheer, dark ravine, picking our way around bluffs. After a while we emerged from shadows to gather on a small point that overlooked the Salmon and the big trees in the bottom. A half mile above us, midslope on the Salmon River breaks, a fire was raging out of control. Great rolling clouds of smoke lofted high and hung back out over the canyon. Somewhere up there were Mitch, Quacks, and Jack. I took out my watch—2:05.

In a short time we began to get cold. We'd been sweating as we hiked, but now we were standing around shivering. The fire lookout relayed to Humphries that Payette dispatch had called and was concerned that we hadn't yet confirmed our head count.

Single file, we started up a rocky gully where we continued the long ascent. In five minutes we had settled into a steady rhythm, plodding up and up and up over one face after another, seeing the tops of the big timber along the river sink far below. I began to feel sick. No one spoke a word. We held up for a minute, drank some water, looked around, spit, wiped the sweat from our brows, and then settled back into the climb again.

This was another of those moments when I had to wonder about smokejumping, or even fighting fire for that matter. It took an hour to come down, and it would be twice that long going up. Even ibuprofen couldn't relieve the burning fatigue in my legs or the painful knee. What it all boiled down to was misery. True, the canyon was beautiful. Beautiful like in a dream, strangely incomphrensible with its grotesque orange sky and old yellow moon. But mostly it was just misery, aching muscles, and more climbing and all of it knowing that the misery might be leading us to the infinitely worse misery of three dead friends. I've heard jumpers talk about "the thresholds of pain." To the degree that you have a high threshold of pain, you're a good man. To the degree that you don't, you're a good man if you act like you do.

Jumpers rarely speak openly about how they handle extreme fatigue, but when they do, they joke about it and claim to be the weakest in the bunch. At such times I just keep my mouth shut. For me, it's always the same. Beyond the fatigue comes the sorrow and with the sorrow comes the loneliness. At the hour of my greatest exhaustion, I am lonely, emotionally frail, and at a loss to do much about it. No matter who claims to be the weakest, in the deep, dark pit of my soul, I know that it is me.

After climbing for what seemed like a very long time, we pulled up next to the bottom of the fire. Moments later, George Steele came down the hill with his group and met us. There wasn't much talking. We took a small break, drank water, turned off our headlamps, and stared out into the darkness. During our climb, the wind had calmed, and the lower portion of the fire had settled down, while the fire up the mountain in the vicinity of the head continued its furious roar. I took out my watch—4:45. I thought I could see light in the east.

"We were right across there," George said, pointing out into the burn. "When it blew up, you guys were across there about two hundred yards above us."

Humphries said, "All right, then. Let's head up and find the end of our line. We'll start our search there."

We moved across the fire area in heavy smoke. The fire was creeping here and there, only intense under the big pines. We picked our way across the hill, all the time coughing and angling up until we reached another rocky point. We held up.

At the edge of the outcropping, a bed of coals covered an area thirty feet in diameter surrounding a large glowing stump hole. Down in a draw to the west, a dark pitchy form was taking final leave of its body of three hundred years, hissing and popping and then vanishing into the night in black smoke.

I thought about what we were about to do—start looking for

something we couldn't stand the thought of finding. I have seen fire fatalities before. The human body burned, twisted into odd positions, smoking, still alive, moving. I've seen the stumps of what once were hands, the dark ash of destroyed genitals, lips gone, the teeth exposed and bloody. I've looked on as a body was covered with a blanket and left to finish dying in the back of a pickup.

"Here it is," Humphries yells down the hill to us. "The end of our line."

We line out across the slope, about fifteen feet apart. The plan is to move up the hill, staying even and gridding the area. Kelton shows us the place where he last saw Mitch and Quacks, and we head up from there. We move slowly, taking time to look thoroughly to both sides. Headlamps flashing through the smoke make for poor visibility, and it is frequently necessary to move back and forth for a closer look at a burnt rock or a charred log. Faint light grows in the east. My emotions oscillate between nervous trepidation and a nearly overwhelming sense of gloom.

Mitch and Quacks. Always together, always fussing and picking on each other. Mitch, the veteran, trying to impart his brand of wisdom to the snookie he fondly called Quacko. And Quacks, hammering back at his would-be mentor, singing his made-up song, "Grandma Mitch, the daffy old bitch."

And Firestone. Jack had already outrun one fire this year at Pothole Lake. Then on our last fire, it was he and Dunning who had outrun the snag. Jack knew fire in the breaks from his years at McCall. But Jack also had a bad knee.

"Wait up," someone down the line yells. We stop. My heart begins to beat faster.

"Everybody stay where you are," Humphries commands.

I turn off my headlamp and stand in the dark watching as Humphries makes his way over to someone. I can see it when Humphries holds it up—a hard hat.

"It's just a hard hat," Steele says.

"Can you tell whose it is?" Bass asks from the darkness.

"Quacks."

I turn my headlamp back on, and we line up again. We move on up the slope into the gray dawn and a haze of blue smoke.

"Hold it," someone down the line says.

I begin to feel sort of crazy. A slight dizziness combines with an absurd need to burst out laughing, or worse, to break down and cry. The line halts once more and Humphries goes over, bends down, and picks something up—the head of a Pulaski, its handle burned off. Lying beside the Pulaski head is a pool of melted plastic, charred debris, and a burned radio—Firestone's PG bag.

We line up again. Thirty yards up the hill Humphries yells something really loud. It was just getting light enough to see our ragged line of searchers. I see John take out his radio and hold it to his face. *Oh no,* I think. *He's found one.*

Seconds later Humphries lets out a weak, muffled yell, then adds slowly and clearly: "They're OK. They've just come out down on the road."

Along the mountainside we let out a few small noises and shut off our lights. The mountain is quiet. I fall to the ground on my knees and elbows and bury my face in my dirty gloves and begin to weep softly. The group keeps apart, each man staying where he is, not moving. After muttering my short prayer into the ashes, I look up. The only sound is the distant murmur of the river and Humphries talking quietly into his radio. Now he can complete the head count he so desperately hoped for. He can have the lookout relay to McCall a message for those who have sat up all night, hovering around the operations desk. All the jumpers on French Creek are OK.

September 10

Lavender clouds sailed like tall ships across a pale green dawn as we headed up the old fire line to our jump spot. No one said a word. Our night on the mountain had left everyone drained.

Somewhere in the fire a snag came crashing down. A shower of sparks rose in the air, and my thoughts drifted back to Mitch, Quacks, and Jack—they were safe now; all of us were going to be fine.

Like us, they had decided that the route up was too slow and had started cutting across the slope, but the tongue of fire that Bass, Kelton, and I had worked our way through was too hot for them to do the same. Caught in a horseshoe of fire, they were forced to try for higher ground. In a panic they happened upon a rocky area of sparse fuels, and there they holed up under their fire shelters until the main front passed by. From there they worked their way to the river, dropping down the same canyon we had. Charlie Brown had radioed from the bottom. Our boys were shaken, especially Quacks, and were being transported to Riggins for treatment of first- and second-degree burns.

Back in the jump spot we found Romo cocooned in a cargo chute, sound asleep next to the gear. All our work had been for nothing. Burned parachutes, cargo chutes, and three mains hung in the trees all around the jump spot.

"I could use some coffee," Bass said, dropping his PG bag to the ground.

"And a jug of whiskey," I added.

We built two campfires, gathered around, and began heating cups of water. Romanello stirred, pulled part of the chute away from his face, and lay staring out, a disheveled-looking thing. We started laughing. We laughed more and more until finally we all were howling, even Romo.

"Quick," Seiler said. "Get a gun. I think it's alive."

Romo crawled out of his bedding.

"That's gratitude for you," he said, still laughing. "I stay up here all night eating smoke and coughing my guts out to save your shit, and all you small-time brush monkeys have to say is 'get a gun.'"

Troop and I sat down together near one of the campfires and

waited for our cups of water to get hot. Bass came and joined us. The rest stayed at the other fire, verbally sparring with Romo.

"How's the hip, Troop?" I asked.

"Oh, it's all right, I guess. It only hurts real bad when I go downhill. Say, I wanted to ask you. You didn't happen to see a letter along the fire line or down on the road anywhere, did you?"

"No, I didn't. Why?"

"Oh," Troop said in a soft, deflated voice. "It was one of my favorites. Kind of a love letter, I guess. Cory wrote it just after we met. I've been carrying it in my PG bag, and once in a while I take it out and read it. I had it just before dark, but I guess it's gone now."

Troop sat staring into the campfire and began talking about his wife and her plan to divorce him. He'd talked to me about it before—a sad footnote to his summer. Bass and I just sat there listening. A few coyotes came over the ridge, yipped hysterically, went silent for a while, then sang out a long, drawn-out farewell to the moon. There on the cold ground, we sat in the smoke, eating beans, drinking coffee, and watching the fire burn in old logs and stumps around the jump spot.

"It's hard to do this job and have a woman at the same time," Bass said.

"Oh," Troop said. "It wasn't hard. She's the one unhappy, not me. She's great, though! Loved everything—except Alaska and my being gone all the time."

Troop finished off his beans, scraping his spoon around the bottom of the can. Suddenly he seemed less blue.

"Ahh," Troop said thoughtfully, "I don't know what it takes to make them happy. But I tell you, when I went to China to talk them into starting a jumper program—that did it. Chinese smokejumpers. That was the last straw for her."

I got up and scouted around for a place to lie down. I found a relatively flat spot a few feet away, removed the bigger rocks, took out my tarp, and put it over my new bed.

The old moon settled in the west. Other guys were digging out places, too. Then quiet. Lying there under a gray and pink sky surrounded by the blue silhouettes of charred trees was a moment of exquisite beauty.

Two hours later I woke facedown in the dirt. Having drooled in my sleep, mud was caked to my lips and the side of my cheek. I sat up stiffly, spitting out sand and ash, and took a moment to clear my head and remember where I was. A yellow sky played across the tree line above where I lay on cold dirt and rocks. I lifted my rain tarp slowly so as not to wake anyone and got to my feet. There was no way I could stay on that ground any longer, cold as it was.

Limping over to our gear pile, I noticed Troop curled up around a pile of ashes. He looked dead. Other bodies were rolled up in parachutes. Most of our sleeping bags had burned. Bass was nowhere in sight. I went quietly to the edge of the clearing, gathered some sticks, kindled a fire, dragged up a log, and sat staring at the heat, still half asleep.

My new campfire crackled cheerfully in a way that made me sort of hurt inside. I got to thinking. Here I am—living the dream, as jumpers like to say. The smokejumper dream, complete with a wholehearted ass-kicking from Mother Nature. The old girl dancing, as usual, to her own tune. And wouldn't you know it, a blues tune at that—the chase 'em, rock 'em, smoke 'em, snake 'em, bug 'em, bear 'em, snag 'em, divorce 'em, sleep-starve 'em, and burn 'em up blues.

The silence of the morning was strange in the aftermath of the chaos of the past sixteen hours. The near entrapment and possible death had left me stunned. This could have been a very different morning. On the hillside below, teams of medical and forensic specialists could be examining several bodies, then covering them with blankets. The supervisor's office for the Payette National Forest could be trying to locate next of kin. Several bodies might have been burned beyond recognition. Dental records would have to be requested. Cover Charlie Brown with a blanket. Cover Troop with one. Cover Bass, Humphries, and Johnson. Pull it over Seiler. And then, finally, cover me so that no one would ever have to see me like that again. The end would have come—grim and violent. How close did we come? How lucky were we to all be together there in the clearing instead of down the hill, separated from each other, each under our own blanket, dead?

Hearing a noise, I looked up to see Troop coming over. I suddenly felt better.

"Beautiful morning, huh?" Troop whispered.

"Yeah. Especially after last night," I said.

Troop smiled to the east as if he were about to receive a sacrament.

"How's your hip?" I asked.

"Oh," he said, his black eyes shining. "She's a little sore. How's the knee?"

"Well, I ain't falling down yet," I said. "But every step sounds like I'm walking on potato chips. You got any coffee?"

Troop lifted his hand to the sky. "We got a whole damn jarful, buddy!"

By midafternoon, we had helicoptered off the jump spot to a sandbar on the river. The wind was up, and the fire was roaring at three thousand acres. The river bottom was different now, busy with trucks, engines, a Class I team moving in, a camp being set up. While we waited for the rest of our gear to be coptered down, we walked to the edge of the river. From a series of rocky narrows, the Salmon spilled white into green pools, then roiled up in expanding patterns of light that drifted off into canyon shadows. A few guys decided to take a swim. One of George's bunch ran to the edge of the water, grabbed his ball sack and dove in, hitting his head on the bottom. He floated to the surface belly-up like a dead fish. We thought he'd broken his neck, but he came to, choking and spitting up water, and everybody had a good laugh.

We loaded our gear into two army trucks and then climbed on top. The trucks maneuvered down the shade-dappled road under big pines, growling slowly past boulder fields and sandy river bottom, then crossed a great high bridge to climb along vertical cliffs. In the back, piled on our gear and on each other, we marveled at the scenery, took our shirts off, let the wind blow our hair, laughed, and carried on like blackbirds in a pie.

21

September 17   The Coast of British Columbia

Peering out the window of the Boeing 737, I look north, wondering how long until we arrive over Alaska. Thrusting up out of the ocean, the Stikine Mountains extend inland as far as I can see.

Aboard the jet, the jumpers returning to Alaska sit together in one area, ordering drinks, telling stories, and flirting with the flight attendants. A couple more hours and we'll be back in Fairbanks.

Our departure from McCall had been sudden and, for most of us, bittersweet. The day after getting back in from the French Creek fiasco, Quacks let Firestone know that he was quitting jumping for good. Standing outside the ready-room doors, his hands and neck wrapped in bandages, he stated it simply:

"This summer's been too much. I can't do it anymore."

Several other jumpers—Mitch in particular—encouraged him to take some time off, think it over. No, he said. He *had* thought it through. French Creek was enough. He couldn't bear the image of some BLM official knocking on the front door of his house and his wife answering it with a baby in her arms, and then standing there while he delivered his message: "I deeply regret that I must inform you . . ."

Quacks pulled his gear off the suit-up racks for the last time. Two days later, back in Fairbanks, he resigned and returned to his family in Juneau. The fire on the breaks had left others shaken as well. Seiler wasn't talking. Troop didn't think it was that big a deal. He remembered Vietnam, saying, "Any day someone isn't shooting at you is a good day."

Mitch, Quacks, and Jack had initially headed up the hill thinking they could make it back to our last safety zone inside the original burn. But when they got there, the interior of the fire was too hot.

"We were going as fast as we could," Jack told us. "Our only option was to outrun the far finger. I figured at the time that it must have gotten the rest of you guys. It became clear that we couldn't make it up around it and surely not through it, so we were forced uphill. Quacks fell down and lost his Pulaski first. A few feet farther, he fell again and lost his hard hat. That's when he panicked. Mitch grabbed him by the shirt and pulled him to his feet. Fifty feet up we came to the rocky point. That's when I yelled, 'Shelter up.' While Mitch and I were grabbing our fire shelters, I looked over and Quacks was just standing there. He just kept yelling over and over 'This is it! This is it!'"

Jack went on. "Mitch and I shook our shelters out side by side on an area of loose rocks and crawled under them. The last thing I remember seeing was a blast of hot air and sparks hitting the point and blowing Quacks, who was still holding on to his shelter, through the air. Then the fire swept over us."

Quacks had somehow hung on to his shelter. The blast blew him ten feet into a crevice between two large boulders, where he was able to wrap the shelter around him. Fortunately, the fuels in the area were sparse and the intense heat subsided after the first two or three minutes.

"I don't know how hot it got under that shelter," Mitch said, "but I just knew that we'd had it. I put my face down into the rocks, and there was air enough to breathe. I never believed in miracles, but I think maybe that was one."

Discussing the incident until quitting time, we each took responsibility for the decisions that had been made. We questioned our judgment in going back downhill after abandoning the line. The fire officer in the helicopter who had radioed us on the fire came and talked with us. He felt terrible but stuck to the notion that his assessment at the time had been sound. We had violated some safety rules. We had taken calculated risks. Too many. Firestone had had his third close call for the season. He, too, was thinking of quitting. A hard thing for Jack after so many years as a jumper. The thought had crossed my mind as well. Smokejumping is a numbers game, and through twenty-six seasons I'd pushed my luck farther than most. If my knee had failed and I'd fallen behind Bass, I would have pushed it one fire too far.

In the three weeks that I had known Carol, the tenderness we had shared had left her confused and me convinced that it was really Sally I wanted. I had come off the French Creek fire exhausted, limping, and mentally traumatized. Part of that evening and the next morning Carol and I spent lost between clean sheets, casting all restraint to the wind. I took my last afternoon off, and we spent it in my hotel room watching country-music TV, sipping beers, and making love while a breeze off Payette Lake rustled the lace curtains and flowed cool over our bodies.

At dark it started to rain, and we took a walk downtown. Sitting at the bar in the yacht club where we'd met, we made small talk and rolled dice for the music. Jumpers came in, got beers, and went to the back to play pool. The smell of rain was sweet as it drifted in through the back doors.

When it came time to go, Carol and I got up and walked outside and stood on the sidewalk. She didn't want to leave, and I considered taking her back to the hotel just one more time. Instead, I gave her one final hug and pushed her gently away.

"It's time," I said. Tears welled up in her eyes. She stepped back, pressed her lips together, turned, and walked away.

Just north of the Alaska Range, the jet idles back its engines for the long descent into Fairbanks. Already I can see the glossy greens of summer turned yellow and orange. The black spruce remain unchanged, filling in vast expanses with dark green, while tundra meadows and ridgetops blend in soft tans. Dwarf birch tint hills and meadows in drifts of pink. Above it all lies a fresh dusting of snow.

At the airport in Fairbanks we load our gear into the box van and head for the standby shack. After a few hours reorganizing my gear and completing time slips, I drive to the barracks. Night has fallen. Rows of tiny blue runway lights stretch from one end of the airfield to the other. A rotating beacon flashes its beam across a featureless night sky. In the north a dim smear of gray-green light appears. It grows in intensity, a cloud of silver, streaking south, then wavering into a silver-green curtain that holds for a moment, slowly extending itself upward into space. The curtain unfurls overhead, undulating slowly, then shudders as highlights of red and gold concentrate along its edges.

September 18

I took the day off and went downtown. On the way I passed the Captain Bartlett Inn and memories of Sally rushed back. While my pickup was being serviced, I walked to a nearby shopping mall and bought Sally some gifts. I thought of calling Sally's brother to see if he had any news of her but decided not to.

The atmosphere back at the shack was quiet and easygoing. Boatner was off his crutches but still limping. Chip Houde had been released from a doctor's care in McCall and had come home with us. Sharp pains in his lower back had Charlie Brown in for X rays. Word had come from the jumpers in Redmond, Oregon, that Tyler Robinson had been released from the hospital and was on his way home. Besides the nine left in Redmond, fifteen others remained in Missoula, but the daily situation report indicated that they would be returning soon as well. In the loft, parachutes were being broken down and sacked up for winter. After working a month on his land near Denali National Park, Don Bell had dropped by the shack to hear a few new down-south jump stories.

I went to the operations desk to fill out a leave slip. Togie sat behind the desk, the rookie fielding phone calls and playing box boy. Suddenly a great uproar came thundering from the paracargo bay, followed by fits of laughter. A few jumpers came out of the lounge and gathered to watch. Joining them, I saw that a new game had just been invented.

During one of the slow times of the previous season, one of the jumpers had made a larger-than-life plywood cutout of a fully-suited smokejumper, painted it, and sawed out the place where the face should be. The idea was that a guy could put his face in the hole and have his photo taken, like people do at carnivals. Someone thought it more fun to have a guy put his face in the hole while others stood back and kicked a soccer ball at it.

Today's kickers had lined up happily thirty feet from the cutout jumper and were drop-kicking the ball straight at the face in the hole— Don Bell's face. The object of the game was to keep your face in the hole and not move no matter how close the ball came, all the while hurling a continuous stream of insults at the kicker. After each kicker had one turn kicking, the kicker at the end of the line was next up for his turn in the hole. Evidently it was more prestigious—although clearly more life threatening—to be the one cursing the mob.

Blam—the ball hit the plywood jumper directly in the stomach.

"That was just practice," Rod Dow yelled to Bell.

"Dow," Bell screamed. "A man who's been known to eat horse-shit sandwiches and kiss a frog's ass." *Blam*—the ball hit the plywood jumper in the right arm, then rebounded back.

Dow handed the ball over to the next kicker, Scotty Dewitz.

"Dewitz," Bell shrieked. "A bald-headed, borderline buffoon. Worse than the idiot Dow. May monkeys fly out your butt and shit in your helmet."

Whoosh—the ball went flying by just a foot over Bell's head.

Romanello lined up and took aim.

"A no-count, weasel-eyed wop," Bell shouted. "A pathetic pizza-pushing descendant of Roman perverts." *Blammm*—the ball slammed into the plywood only inches from Don's face. He blinked his eyes but kept yelling. "Wine-guzzling greaseballs, spaghetti-sucking . . ." *Blam.*

The next thing I knew, guys were running up with more soccer balls and basketballs. A barrage of balls went slamming into the plywood, but Bell hung in there cussing a storm.

"No good, gutless, good for nothing . . ."
Blam. Blam. Blam.
"Worthless scumbags, worms, low rents . . ."
Blamm. Blamm. Blamm.
"Weak-tit, mouth-breathing, pinko sleezebags . . ."
Bell dropped the plywood cutout and started grabbing balls. I rushed in to help, and we starting throwing balls and chasing jumpers in all directions.

Early that evening after a quick good-bye to a few guys, I climbed in the pickup, turned up the stereo, and roared out of Fort Wainwright, leaving its sleepy blue runway lights in the dark behind me.

I stopped for a cup of coffee in Big Delta and continued on south toward the Alaska Range. At midnight I reached Paxson Lake in a rainstorm, pulled over, and slept three hours in the front seat with all my stuff. I woke up shivering at three in the morning.

A light snow covered the windshield. I started the pickup and headed down the road toward Glenallen and lower elevations, where the snow turned to rain. By seven o'clock I was passing Chitna. A few minutes later I crossed the Copper River and hit the old railroad grade.

I counted off the mileposts, twenty, forty, then fifty-five, and finally I caught a distant view of McCarthy across the valley, ten miles away. The sun was up bright now as I traveled the straight stretch to the river. Rattling and bumping, I emerged from the lane of birch into the full light of the parking lot. Before me was the river, the cable crossing, and only a mile away, the shiny metal roofs of McCarthy.

I made the two river crossings, noting that the rivers were even more powerful than at midsummer. In the shade of the tall aspens, I walked into town and up to the lodge. A few people sat out front on wooden benches, reading and talking. Inside, sitting at the same table as two months before, were Denny and her husband.

"Hello," I said.

"Hello, can we help you?"

"I'm a friend of Sally's. Is she here?"

They looked at each other for a moment.

"Oh, you were here before, weren't you?" Denny said. "You missed Sally."

My heart stopped.

"Yes, you missed her. She went horseback riding just a half hour ago. She'll be back after a while."

"Do you know which way she went?" I asked.

"Not exactly," Denny smiled pleasantly. "But she often rides up by the glacier, especially when she gets a late start."

I asked if I could leave my pack in a corner, then walked out in the street and looked both ways. I stood around trying to look casual, wondering what to do. I thought about going after her, but then I might miss her return.

After waiting for two long hours, I looked up and there was Sally riding into town, wearing her Celtics ball cap and the checkered brown-and-white flannel shirt I'd given her on my first visit. A young girl, nine or ten years old, rode beside her. Sally looked wonderful, her arms tan and the tips of her hair bleached blond.

She pulled the horse up to a hitching rail in front of one of the buildings. I watched as she got off and tied the reins. I walked up behind her as she was saying something to the girl. She turned.

Her mouth dropped open; then a serious look crossed her face.

"What are you doing here?" she asked.

"Well," I stammered, "I just got back from Idaho, and I missed you . . . so I thought I'd come."

"Didn't you get my letter? I wrote you asking you not to."

Somewhere deep inside me, a white stone disappeared in a black pool.

"I've come a long way," I said. "I had to see you."

"My brother's coming down this weekend, and I wanted to spend time with him."

Sally pulled off the saddle and turned and looked at me severely.

"I don't have time to talk now, I'm running late for work. We'll talk later." She reached out and touched my arm. "Murry, I'm sorry, but things aren't the same with us anymore."

Sally untied the horse, turned, and led it around the corner of the building. Standing there stunned, a wash of sorrow welled up in my chest. I'd tried to warn myself not to hold on to any particular expectations with Sally. I'd told myself that it was best to be ready for anything. But I hadn't listened. It all became clear how much I'd counted on the moment when I could come back for her, and she would say yes to my proposal to share winter together in the mountains of Northern California. In my most difficult and trying times in Idaho, I had imagined the joy of returning to Sally and used it to get me through.

I took a short walk, then returned to the lodge, ordered a beer from the bar, and took a seat at the grand piano in the rear, close to the doors that led to the kitchen. Halfway through the second piece Sally came over in her working dress, looking sad.

"I'd forgotten how lovely you play," she said, tucking behind her ear a lock of hair that immediately fell out again.

"And I'd forgotten how lovely you are."

"Murry," she said. "You're a wonderful man, and you have a very interesting life. But you weren't *here* this summer. You were where you always are in summer. It's you, it's your life, and it's a fine one. But us being apart brought out some intense needs in you. You need something more than what I have to give right now. I was thrilled at first, but then, while you were gone, I began to wonder. Your last two letters were just too much. I know that you want to be in love, but I'm just not ready."

I kept playing and looking at Sally. She reached out and touched my shoulder.

"Over the Labor Day holiday I got involved with someone here . . . We've been seeing each other." She stopped and forced a smile. "Murry," she said, pausing for a long moment. "I don't expect much to come of it. It was just here and you weren't. I'm sorry. I can't talk now. I've got to get back to work."

Sally turned down the corners of her mouth, tipped her head side to side, and shrugged.

"I brought some things for you," I said, not looking up from the piano. "They're in the pickup. I'll get them and then leave in the morning."

Sally disappeared through the swinging doors of the kitchen just as I finished the last piece. I left the lodge and walked to the river.

At last I had my answer but not the one I'd come for. Not only would Sally not be spending time with me traveling or wintering in my log cabin, Sally wasn't going to be spending time doing anything with me, ever. As unbearable as it was, I had to admit down deep that I would most likely never see her again.

After checking into the old Johnson Hotel across the street, I went for dinner at the lodge. Sally brought me coffee, paying me a little special attention. She insisted on treating me to a free meal. After dinner I went to the bar, bolted down two shots of whiskey, and shot a game of pool by myself. Then I told Sally the number of my room and left. It seemed like an eternity before I finally heard her footsteps on the porch. I'd waited in the room all evening, staring at the ceiling, feeling empty and lost, and needing it all to end.

She tapped on the door, and I opened it.

"Hi," I said.

"Hi there." Sally smiled.

I stood aside and allowed her in.

"Sally, I brought these things for you from Fairbanks. I don't know if you need them, but I thought . . . being way down here . . ."

"Oh, Murry. You've always been so thoughtful."

Sally opened her presents—film, another Enya tape, orange-scented bath soaps, a silk blouse and green sweater with a gold pig pinned on the front.

"These are lovely. I don't know what to say. Thank you! And thank you for everything. You're a special part of everything my summer's been."

"Someday, Murry," she said, looking at me and taking my hand, "you'll find what you're looking for. Believe me, there are lots of women who would love a man to shop for them and be like you."

"I guess I let myself get carried away," I told her. "In a way I think I fell in love the first time I saw you. Whatever it was, it was strong for me, and I put a lot of hope in it. Things were rough down south, and I guess that made me needier than I should've been." Tears came and I felt embarrassed. We held each other and rocked back and forth.

"I don't know," Sally said softly. "I was thinking of traveling with you. I missed you, too, and wanted to be with you again and make love like we did. But then, I met . . . this person and things just sort of happened. I'm not serious about him, but since it's happened I've realized that I need to take some time and figure out where I'm going with my life. A lot of my past has been me just falling into situations and sort of accepting them rather than thinking out what I want ahead of time."

We talked a little longer, then Sally stood up. "I have to be going," she said. I stood up. We had a long hug. Then Sally turned, opened the door, stepped through it, and pulled it tight behind her.

September 20

The room was cool when I awoke at 5:30. I immediately got out of bed, packed my day pack, left the key on the dresser, and stepped outside. Raindrops hung like crystals on every branch, leaf, and twig. A storm had brought down a new batch of aspen leaves and scattered them, bright yellow, upon the dark brown earth. Walking out into the street I took one last look at McCarthy and the little window of Sally's room. The sky was blue and mostly clear.

September 24

Returning to the standby shack was hard. I spent one day just looking out the windows, watching the sky grow dark and trying to hide the fact that I was an emotional mess. My pride had been wounded. I felt embarrassed. Keeping to myself like a sick cat, I locked my room at night, pretending not to be there when someone knocked. I quit eating. I wasn't sleeping well. When I did manage to sleep, it was fitful and tormented by weird, unsettling dreams. Within moments of waking, I became swallowed up again by a terrible sinking feeling.

Most all the jumpers were aware of my involvement with Sally—a few had cautioned against it. They also knew what had happened in McCarthy. From time to time Troop, Dow, or Charlie Brown checked on me.

"Wanna be my Nutty Buddy?" they'd ask. We'd hit the freezer, then go outside, take a seat on a bench next to the ramp, and look out across the runways and talk about how the season had been, how many good fires we'd had, the big money we'd made, and how lucky we were to have such a great job. Each tried in his own way to cheer me up, but I was so caught up in feeling sorry for myself that even what little laughter we could manage left me sadder. On top of it all, I was thinking of quitting jumping.

During break time this afternoon I walked out in the yard to the end of the sidewalk and sat down on a picnic table. Before we left to go down south, some of the jumpers had begun landscaping the new yard. One had picked up a sculpture at a local flea market and placed it in a flower bed that bordered the main sidewalk near where I was sitting. Two feet tall and almost that wide, the rough rectangular piece of granite was eight inches thick and had the appearance of a tombstone. Near its center was a woman's face, turned slightly to the side, chin lowered, eyes cast down. Not a sad face actually, but one that would never again look out upon the world; a fitting testimony to the many fine women who smokejumpers had lost through the years.

Back inside the shack, there was a letter for me.

September 7

Dear Murry,

Got your letter and postcard on Wednesday. It's been quite a steady stream; you've been so generous. Thanks for keeping in touch. I do need to inform you of a few things so that you can make plans accordingly. I'd advise you <u>not</u> to come down after you get back from Idaho. I'll be working (i.e., long days, extra hours, <u>very</u> busy) the entire time. Secondly, my brother is coming down one of the weekends and whatever time I have, I want to spend with him. It's just not a good time to visit. Third, please don't plan on picking me up here. My plans change like the wind, and I'm tentatively due to stay here until the 20th. It might be longer, who knows.

Basically I'm telling you to go ahead with your plans—don't try to build them around me. If you weren't planning on coming back to Alaska, except for me, <u>please</u> don't. I'm being up front. Once I leave here, I don't know what I'm going to do. But I want it to be open, flexible. I don't want to make plans that lock me in. I've got a lot of things I'm considering doing but can't exactly check them out while I'm here.

I like you, Murry, and I'd like to see you again. But I'm not looking for anything—or anybody—to tie me down.

Love, Sally

That evening I sat in my room and watched out the window as Chip Houde made several trips back and forth between his room and his pickup, loading up for his trip south. As always, that's the way it would end. After our summer of intense togetherness, for the most part each of us would be going our own way privately rather than saying our good-byes in person. Just hurry up and load and go.

There came a rough knock at the door. I thought about not answering.

"It's Troop, dammit. Now open up."

I opened the door and in stepped Troop carrying a twelve-pack and a huge bag of Fritos. "Hey, old man. How's life?"

"Oh," I said, "it's all right. I'm a little blue."

"Yeah, I know," he said. "Me, too. Ain't it a pisser? Work hard all season and wind up going home to a mad woman."

I sat looking at Troop while he opened us a couple beers and tore into the bag of Fritos.

"It's got me down," I said. "I think maybe I've had enough."

"Enough jumping?" Troop asked. "You can't be serious."

"I don't know," I said. "This summer took a lot out of me. I lost twelve pounds, I can't sleep, my knee's banged up, my head's a wreck, and now everybody's leaving."

"Well, sure," Troop said, "but what's new about that? You'll get over it. Just be glad your girlfriend isn't asking for fifty thousand dollars so she can go to college. That's what my wife wants."

"Damn, that's a lot of overtime," I said. "Seems like she'd be happy enough just to get rid of you."

We had a good laugh at that, and I popped open two more beers. There came another knock and Mitch, Togie, and Bell walked in. They also had beer.

"Watch out, Troop," I said. "It's the Ladies' Aid Society in drag."

"Yeah, right," Mitch said, "and you're the world's top expert on women."

Bell and Togie went down the hall and brought back some chairs.

"We just figured Old Leathersack had hung out licking his wounds long enough," Mitch said. "And look who's here, another old fart flailing in the quicksands of love. Troop and Sack, the Laurel and Hardy of romance."

"Me?" Troop wailed. "What'd I do? I was happily married. I didn't want a divorce."

"The old fools are always gabbin' about women," Togie said. "Old fools" was something the rookie had come up with in referring to a few of us older jumpers. He'd spread it around, and now it was a favorite saying on the crew. Everybody liked Togie. He'd had a great year, and he knew it, so he'd taken up the habit of naming jumpers all kinds of silly nicknames.

"I may be a fool," Bell protested, "but I'm not old, so don't go putting me in with the old fools."

"Hell," I said. "It's worse to be a fool than old. A guy can't help being old."

"Bell can't help being a fool, either," Mitch said. "So he belongs with the other half of the crew—the young fools."

"Don't forget the complete idiots," Togie said. "That's the biggest group."

We laughed and pulled on our beers. The door opened again and Rod Dow stepped in.

"Another old fool," Togie said. "I knew the geese were gathering up to go south, but I didn't know the old fools did that, too."

I reached in the fridge and got Rod a beer.

Dow smiled. "What are you guys doing, trying to rescue Taylor from himself?"

I quickly diverted the conversation to winter plans. Mitch was leaving the next day for the coast of Spain to help organize a world parachuting competition. In a couple weeks, Togie was quitting for the season and moving to Anchorage to spend his winter attending diesel mechanics school. Bell wasn't sure exactly where he was going, but it would be south for sure, probably somewhere on the Oregon coast. Troop would also return to the coast of Oregon. Dow would go home to Prosser, Washington, and spend his winter writing grants and conducting research on a national training plan for schoolteachers.

When they asked me about plans, I told them I thought I'd spend my winter at home in the mountains and maybe write a book about jumping.

Mitch asked, "What are you going to call it, *Old Men and Fire?*"

"No!" Togie insisted. "*Old Fools and Fire.*" We laughed some more.

"Don't forget to put in the part where the mosquitoes bit you on the balls," Bell roared.

"And don't forget to put in that you spent the summer so worked up over that little redhead that you could barely screw hose together," Mitch added.

We reminisced about Togie's first fire jump, bailing out over his village, and the hard fight on the Clear fire. We talked of the Brooks Range, the Border fire, and the Big Flip. In the midst of it all, I found myself thinking of Sally, and how perfect the summer would have been if things had worked out differently. We recalled many days and nights out in the wilderness, working endless miles of fire, hiking ridges at dawn, and some very pleasant evenings cooking around campfires, telling tales, and living in the bush.

We spoke briefly of Rick Robbins, how he'd broken his back during rookie training and ended his career before it even started. We reflected a little on Billy Martin's death and the near tragedy we'd almost suffered at French Creek. And Quacks, in losing him, we'd lost a good man.

Sometime about eleven o'clock the beer ran out, and we said our good nights.

In the quiet that followed, I sat on the edge of my bed, and watched out the window as Chip Houde closed the back of his camper, got in the truck, and drove away, his red taillights growing smaller, reflecting on the wet pavement, then disappearing around a turn.

The next morning after PT, I was standing at the box staring at the jump list, while the Rogue Togue sat behind the ops desk nursing a cup of coffee and thumbing through a *Playboy* magazine. Charlie Brown came up behind me. At roll call he'd announced that he'd be leaving right after PT.

"Hey, Murr," he said. "How's my old saw partner doing?"

"Char Barr, amazing cat."

"We've got to talk," he said, gesturing toward the hall that led to the paracargo bay. I said, "What'd I do now?"

"You fucked up, Murr, you been fuckin' up real bad," he chuckled.

Inside the bay we stood behind a pile of fire packs. A couple of jumpers were at the other end of the building throwing darts. Bob Dylan was singing "Knock, knock, knockin' on heaven's door..." Charlie turned to me, his eyes dark and calm, the scars on his neck still red from the saw dogs of French Creek.

"Murrman," he said, "it's been real." He put out his hand and we shook on it.

"This has been a great season for me," Charlie said. "My best ever."

"Well," I said, "you've got the scars to prove it."

Charlie reached up, touched his neck.

"A couple scars, Murr, is a small price for what we got to see."

"I guess in that way, it's a lot like love isn't it?"

"It's not *like* love. It *is* love. It'll take time before it all sinks in; so much happened in so short a time. Not everybody gets as close to life as we do."

"Not everybody would want to," I said. "The whole thing has kind of overwhelmed me."

"Your Sally was the only thing that overwhelmed you," he said. "And you brought that on yourself. You got lost in it. You were more in love with love than you were with her."

I looked at Charlie for a moment. Busted again by the love police.

"Do you charge for this kind of information or is it free?" I asked.

"I haven't got the heart to charge knuckleheads like you. I just wanted you to hear it from a friend. You're going home sad, and I don't think you need to. You've had a great season, and you're walking away with the blues."

"Listen, Oprah," I said, "it wasn't exactly a relationship of convenience, you know. It hurts, that's all."

"I'm no Oprah, Murr. I'm just your friend. And I'm saying you need to take a look at it. You're a lucky man."

Tears welled up in my eyes, and I had to look away.

"Not everybody's got a heart like you," he said. "I know it gives you pain sometimes, but think what it gives in return. Not everybody feels things as strongly as you do. You hear me, Murr? There's two sides to it. Now I've got to get going."

We glanced at each other briefly.

"Think about it, Murr, that's all I'm saying."

We shook hands, and Charlie walked off.

A few minutes later Mitch walked up to the box with a day pack slung over his shoulder and stood there nonchalantly, making it clear to everyone within earshot that he was exceedingly pleased to be leaving us.

"I'm glad I'm getting out of this loony bin," he quipped. "I'm not used to spending this much time working with the mentally challenged."

"You're not used to spending this much time working, period," Bell said. "Now, go on and get your ass out of here."

A quick handshake and Mitch hit the door. "Good-bye, worms," he said. "See you next spring."

That afternoon I talked with Boatner and decided to make it my last. I spent the rest of the day checking in my jump gear and signing out.

September 26

A west wind blew hard during the night, scattering the leaves of summer and bringing with it a cold rain. In the dim half-light of dawn Rod Dow came to my room to help load my gear and take me to Fairbanks International.

As we drove out past the end of the main runway, I looked out across the airfield. Rod was looking, too.

"Strange, isn't it?" he said. "This saying good-bye."

For a moment the airfield was no longer dark but appeared as it had in early July, expansive and sprawling beneath a gray, smoke-filled sky lit by a solitary red sun. Down the main runway *Jump 17* came rushing, its sleek beige form smartly trimmed in red and black, to lift its nose, take to the air, and sweep up over the Chena River and climb north over town. "Fairbanks dispatch, this is *Jump 17*. We're off the ground at this time with twelve souls on board, three and a half hours' fuel, enroute to the Squirrel River fire . . ."

I looked back at Rod and smiled.

"Strange, indeed," I answered. "Maybe someday I'll be able to do it."

The windshield wipers worked back and forth as classical music played softly on the radio. The cab felt cozy and warm. Between the railroad tracks and the river, a group of Canada geese suddenly lifted off the ground. We rolled to a stop, cranked down the windows, and listened to them calling. After flying a wide pass around us, they strung out in a tattered V, caught the wind, and headed south under low-flying clouds.

Epilogue

December, 1999   Quartz Valley, California

After two days of rain, the sky has cleared. As I stand on the front porch of my cabin, looking out over Quartz Valley, the sound of rushing water filters up through the trees from the creeks below. An expanse of valley floor runs north across ranches, grasslands, and fields to disappear beneath a layer of fog that has formed along the Scott River. Stringers of pine border pastures and meadows. From the dark shadows of Blue Pond up into snowfields of Big Meadow, an early morning rainbow arches against the Marble Mountains.

I load my arms with firewood, step inside, stack it by the stove, then put a pot of water on for coffee. In the cozy warmth of the cabin, I feel rested and strong.

Searching through my files a few days ago, I came upon Sally's letters and two sketches she'd sent me at the end of the 1991 fire season.

Bozeman, Montana
November 15

Dear Murry,
I hope this finds you well and happy back in California. I'm here for a few days visiting an old friend of the family's. I left

McCarthy two weeks after you were there and went to Anchorage to stay with Denny's family—the ones who own the lodge. I'm on my way to Pennsylvania now.

I know the summer didn't turn out like you wanted, but for me, meeting you was wonderful. You opened a window into a couple things I've always wanted to believe in—<u>Me,</u> and the idea that <u>Life</u> is an <u>Adventure</u> and was meant to be lived that way. Now that that's done, I need to put some of my newfound wisdom into practice. That will best happen being free from entanglements for a while. I could love you very easily, you know. And maybe there will come a time for that. But for now, I have other business. Here's a sketch I did for you.

Love, Sally

I spread out the sketch on the kitchen table and placed books on its corners to hold it in place. The date: September 21, 1991—the day I left McCarthy. The letter and sketch were the last I ever heard from Sally. I rarely think of her anymore, but when I do, I think about how much I'd still like to have a love story in my life. That part of the puzzle remains unchanged. Since the summer of '91, I've been involved with three different women. One, I was deeply in love with. Another story, same ending. My winters I spend mostly alone.

Along with the accumulated sorrows of lost love has grown an increasing appreciation for the life I'm able to live as a smokejumper: jumping fires during the summer and spending the rest of the year working my land, reading, writing, and traveling the world.

The summer of 1991 had come to an end. As it turned out, it had not been an easy one, but a great one instead. The crew had returned from Idaho tired, but triumphant—even Quacks. Despite his calling it quits, he'd come home proud. No one made any judgments about his choice. For me, it would take until the end of October until I was eating and sleeping normally. In the meantime, my knee mended.

I returned for the season of '92 and the following three as well. By '95 I had become the oldest active jumper in the country. During a training run in the spring of 1996, a final pain shot through my left knee, and it collapsed on me. On May 25, Dr. Keller performed arthroscopic surgery, trimming the torn edges of both meniscus pads, smoothing the kneecap surface, and removing fragments of torn tissue and damaged cartilage.

I was running again by July 4. But it was too far in the season for jumper training. Three weeks later I strapped on my new knee brace and volunteered for an assignment as crew boss with an Eskimo crew from Pilot Station. We spent the month of September in the rugged smoke-filled mountains of southern Oregon, fighting six fires just west of Crater Lake National Park.

In spring of the 1997 season, I returned to jumping. By September, with the other knee acting funny, I began orienting my workouts toward strengthening the muscles that supported my knees. In April of 1998, I returned again for another season. By midsummer, I'd become the oldest smokejumper in history. In July of 1999, I made my two hundredth fire jump near Rampart. Al "Togie" Wiehl was on the same planeload and jumped his one hundredth just eight miles downriver from where he'd made his first.

Perhaps being the oldest smokejumper in its sixty-year history isn't that much of an achievement, but to me it means a lot. It's been my good fortune to know a lot of smokejumpers. I know who they were, and I know what they stood for, and I saw what they did. That my name will sometimes be mentioned in the same company as theirs is an honor.

Looking back on some of those long, tough seasons, it's hard to accept that many of the old-timers are gone—Beltran, Bell, Boatner, Firestone, Troop. Don Bell left jumping after the season of 1992 to become a BLM loadmaster on one of its cargo ships. He spends his winters exploring the Oregon coast and living in an apartment on a rocky cliff, high above the Pacific, near the small community of Oceanside. Two years ago Don became a member of the Fairbanks Symphony, playing the kettledrums. Erik the Blak left jumping after the season of 1995, retiring to eastern Oregon. Al Seiler has jumped these last few seasons wounded. Unable to finish the 1998 season, in early August he took a temporary detail as an assistant fire management officer with the BLM in eastern Montana. In October, he returned to Alaska and had radical surgery on his left knee, restructuring two lateral ligaments, and replacing his anterior cruciate ligament. He returned in 1999, still as tough as ever. In 1997, Tom Boatner was promoted to State Fire Management Officer for Montana. In the spring of '98, Dalan Romero became our new base manager.

Troop left smokejumping in '93, married a young woman from the Philippines in '96, and these days he and his wife are the proud parents of a beautiful three-year-old boy. Mitch also married—an elegant, skydiving Spaniard with dark eyes and long straight hair. Kubichek gave up on becoming a commercial pilot and returned to jumping full-time, got married, and he, too, has two lovely children. Even Dow got married. He and his wife have a fine two-month-old son.

Jack Firestone left jumping after 1992 and now works as lead Air Attack supervisor. Buck Nelson has continued to jump every season. In late August of '98, however, on a jump in the Bob Marshall Wilderness Complex, Buck compound-fractured his left leg. He spent part of the winter at his folks' farm in Minnesota, healing. Last spring he returned to Alaska and had a very satisfying season jumping in both Alaska and the Lower 48.

A few others have been less fortunate. The week after Quacks quit smokejumping, he was accepted into the training academy for the Juneau Police Department. Three months later he was dead. Killed in the line of duty, having fallen to his death while attempting to rescue an elderly man from the side of a building in downtown Juneau.

On September 21, 1998, seven other Alaska jumpers and I jumped the Cave Point fire in the Clearwater National Forest in central Idaho. Togie was there—his eighth season. It was a day that had everything smokejumpers live for—a hot one-acre fire, steep country, tall trees, a challenging jump spot, and a breathtaking view of a sparkling bend in the Clearwater River. It was a Friday and a day any firefighter would be pleased to live, with its demand for quick action digging hot line, eating smoke, falling snags, and relying on one another's skills to catch a tough little fire with potential to go big.

For me, it will always be a day remembered in honor of someone who wasn't there. For it was on that Friday, a hundred miles to the north, somewhere near Libby, Montana, that our friend and ex-jumper Gary "Secret Squirrel" Dunning fell to his death while rock climbing. Gary's life passion had been climbing—both rock and ice. On the day he died, he was free climbing alone, something he'd been doing a lot lately. Recently, he'd bought himself a new mountain bike and was baptized into the local Methodist church. When I asked one of our jumpers who had talked with the rescue crew how far Gary had fallen, all J. L. had to say was, "Far enough . . . He died instantly."

Five months later, I received word that Paul "Papa" Sulinski had been killed in an automobile accident near Willits, California.

It's the first day of December, and another year will soon come to an end. Snow shows on the slopes just above the cabin, and my days are filled with reading, writing, and deciding whether to jump again. When the training seems too hard, I picture the faces of my friends, listen to their voices, and concentrate on how much I still want to be a smokejumper. I imagine Alaska with its broad pastel skies, magnificent sweeping valleys, lines of thunderheads building over the Brooks Range, woolly mammoth tusks curling out of dark riverbanks, and soft pink dawns framed within the curl of lichen-covered caribou antlers.

I am growing old. Each year the end becomes harder. I no longer merely say good-bye to a season but understand that soon I'll be saying good-bye to jumping once and for all. Ten years from now Troop, Dow, Mitch, Seiler, and the rest of us will remain together only in pictures hanging on walls.

Perhaps the day will come when I will no longer be able to remember their names. Their strong, laughing faces, I shall never forget.

ALASKA SMOKEJUMPERS—1991

Sandy Alstrom	Wayland, Massachusetts
Paul Bannister	Swarthmore, Pennsylvania
George Battaglia	Atlanta, Georgia
Gary Baumgartner	Fairbanks, Alaska
Tony Beltran	Fairbanks, Alaska
Allen Biller	Timberville, Virginia
Tom Boatner	Fairbanks, Alaska
Charlie Brown	Rancho de Taos, New Mexico
Lance Clouser	Fairbanks, Alaska
Ken Coe	North Pole, Alaska
Dan Crain	Las Vegas, Nevada
Clay Dalan	Fairbanks, Alaska
Mitch Decoteau	Yellville, Arkansas
Scott Dewitz	Ester, Alaska
Rod Dow	Prosser, Washington
John Dube	Manzanita, Oregon
Gary Dunning	Fairbanks, Alaska
Hank Falcon	Clovis, California
Bill Ferguson	Santa Barbara, CA
Jack Firestone	Fairbanks, Alaska
John Gould	Fairbanks, Alaska
Dave Hade	Fairbanks, Alaska
Kent Hamilton	Orinda, California
C. R. Holder	Missoula, Montana
Chip Houde	Morro Bay, California
Wally Humphries	Sacramento, California
Mike Ierien	Carson City, Nevada
Jim Kelton	Littleton, Colorado
Jim Kitchen	Terlingua, Texas
Tom Kubichek	Shepard, Montana
John Larson	Medford, Oregon
Troop Lugtenaar	Nehalem, Oregon

John Lyons	Moscow, Idaho
John McColgan	Fort Collins, Colorado
Bert Mitman	Fairbanks, Alaska
Bruce Nelson	Fairbanks, Alaska
Jim Olson	Sandpoint, Idaho
Tony Pastro	Fairbanks, Alaska
Eric Pyne	Fairbanks, Alaska
Bob Quillin	College, Alaska
Jim Raudenbush	Fairbanks, Alaska
Dave Righetti	San Luis Obispo, California
Rick Robbins	Canon City, Colorado
Tyler Robinson	Missoula, Montana
Tom Romanello	Whitefish, Montana
Dalan Romero	Taos, New Mexico
Rene Romero	Taos, New Mexico
Rick Russell	Redding, California
Eric Schoenfeld	Haines, Oregon
Al Seiler	Glendora, California
Chris Silks	Gardiner, Colorado
Paul Sulinski	San Carlos, California
Doug Swantner	Meadow Valley, California
Murry Taylor	Greenview, California
Mel Tenneson	Trout Lake, Washington
Steve Theisen	Trout Lake, Washington
Rick Thompson	Mount Baldy, California
Mike Tupper	Las Vegas, Nevada
Rodger Vorce	North Pole, Alaska
Leonard Wehking	Salt Lake City, Utah
Al Wiehl	Rampart, Alaska
Brent Woffinden	Orem, Utah
Chris Woods	Lompoc, California

Names not listed for the 165 smokejumpers of the booster crew that supported Alaska during the main six weeks of activity.

Glossary

Air Attack The planes and people that coordinate air operations over a fire.

Air tankers Large aircraft that drop fire-retardant chemicals.

Backfire Fire set to purposely influence the direction or rate of fire spread.

BIFC Boise Interagency Fire Center. Boise, Idaho.

Big Ernie The smokejumper god. A deity with a rather twisted sense of humor, justice, and fair play. Determines good and bad deals for jumpers.

Blowup Catastrophic fire behavior, rapid spread, mass ignition of large areas.

Buddy check Last-minute check of jumper's gear, performed by jump partner prior to jumping.

Burnout Fire set to burn areas between control lines and main fire; denies main fire of fuel. A tactic used once control lines are established.

Bush General term for the Alaskan wilderness.

Bust	Intense period of lightning fire activity.
Cat line	Fire line constructed with crawler tractors (bulldozers).
Cutaway clutch	The handle used to cut away from a malfunctioned main canopy. Also called the clutch.
Contained	A fire is contained when its spread has been halted by control lines or natural barriers.
Controlled	A fire is controlled when enough work has been done to insure it will not escape.
Crown fire	A fire burning hot enough to continuously spread through the tops of trees.
Demobe	Short for demobilization. The action of leaving a fire once it is out.
Drift streamers	Weighted pieces of colored crepe paper used to determine wind drift before jumping a fire.
Drogue	The small parachute that first stabilizes jumpers as they fall from the plane then pulls the main canopy out of the deployment bag once the drogue release handle is pulled.
Drogue release handle	*See above.* Once widely known as the rip cord.
EMT	Emergency Medical Technician.
Extended attack	Work done after the initial effort has failed to stop a fire. For jumpers, usually the second or third day.
Fat boy box	A cardboard box that comes in the fire packs and contains packaged and canned goods. Jumper rations for the first three days.
Firebrands	Large embers or chucks of burning, airborne material.
Fire devil	Whirlwind of fire.
Firestorm	A mass conflagration of fire, a blowup.
Flanks (of a fire)	The side boundaries of a fire looking from the tail toward the head.

Fusee	Railroad flares used to light burnouts and backfires.
Head	Hottest and most active part of a fire; determines the direction the fire is moving.
Helitorch	A firing device on a helicopter, which is capable of starting fires.
Hootch	Sleeping arrangement: tent, rain fly, parachute, etc.
Hotshots	Organized fire crews; highly motivated and well trained. Mostly used on large, long-term fires.
Initial attack	First effort to stop a fire.
Jump list	A rotating list that determines the order in which jumpers are assigned to fires.
Jump ship	Smokejumper aircraft.
Jump spot	Designated landing area.
Loft	Room where chutes are rigged and maintained.
Lower 48	Alaska talk for the contiguous United States.
Moose-eyed	Jumper talk for being in love. Being *moosey,* feeling *moosey,* having *moose eyes,* etc.
Mop-up	Final stage of fire fighting—digging up all roots and burning material; putting out the last of all embers and coals.
Mud	Aerial fire retardant dropped by aircraft. Also called *retardant* or *slurry.*
On final	For aircraft, the final flight path before jumpers jump. For jumpers, the final flight path as they descend into a jump spot.
Ops	Operations desk. The nerve center of any smokejumper base. The jump list, aircraft list, and all other matters of business are managed in operations.

Paracargo	As a group, those who work to deliver supplies to fires by aircraft and parachute.
	As a product, supplies delivered in such a manner.
PG bag	Personal-gear bag.
PT	Physical training. As part of their regular daily routine, smokejumpers do one hour of PT each morning.
Pulaski	Fire-fighting tool. An ax with a grub hoe on the opposite end.
Rat-holing (also ratting)	Sneaking prized food items and hiding them from the rest of the group.
Rats	Army rations, C-rations, MREs (Meals-Ready-to-Eat).
Ready room	Room in smokejumper facility where jumpers suit up for departure to fires.
Reburn	A fire that is declared out, then later rekindles.
Retardant	See *Mud*.
Rookie	First-year smokejumper.
Scratch line	Minimal hand line, made quickly to temporarily hold a fire until the line can be finished.
Situation report	Daily report of current fires, personnel assigned, and resource allocations. Also includes weather forecasts.
Slash	Debris left after logging; limbs, cull logs, treetops, and stumps. Can also be natural forest debris.
Slopover	A place where the fire crosses an established control line.
Snag	A dead tree, still standing.
Snookie	Second-year smokejumper.
Speed racks	Racks on which jump gear is pre-positioned to facilitate fast suit up.

Spot fire	Fire started outside the main fire area by flying sparks or embers.
Spotter	Person who directs the jumping from the plane.
Spruce bough	The top cut from a small (four- to five-inch diameter) black spruce. Used to swat down flames on Alaska fires.
Stall	In aircraft, when the airspeed gets so slow that it can no longer maintain flight attitude and begins to fall.

In a square parachute, when the canopy is slowed down so much that it can no longer maintain flight, and it begins to rock forward and back radically. |
Standby shack	The main smokejumper building. Includes loft, ready room, tool room, weight room, paracargo bay, etc.
Steering lines	The right and left lines used to steer a parachute.
Stevens connection	A short nylon line that connects the reserve deployment handle to the left riser of the main parachute. When a main malfunctions and has to be cut away, the Stevens automatic pulls the reserve handle and initiates the deployment of the reserve.
Streamer	Fully malfunctioned parachute.
Tail	The back end or initial part of a fire. Usually spreads slowly at lower intensity than flanks or head.
Zulies	Missoula smokejumpers

Acknowledgments

I am deeply grateful to the many people who shared their time and expertise in creating *Jumping Fire*. Like all adventures, the process of writing this book put me in touch with people more wonderful than I could have imagined. From April of 1991 until February of 2000, the effort took ten-thousand hours. The learning curve was steep and constant and challenging. In those dark hours when I struggled, what kept me going was the conviction that I had a good story, and that if I didn't give up, maybe it would be published someday. That would not have happened without the help of many friends, family members, and colleagues.

I thank Buck Nelson, Rod Dow, Chip Houde, Tom Boatner, Bob Quillin, and Kim Field for reading the manuscript in its early form. Thanks to Davis Perkins for his maps, artwork, and photos. Thanks to Chris Woods, Dennis Terry, Steve Nemore, the BLM, Craig Irvine, and Mike McMillan for the use of their photos. (You can see more of Mike's work at www.spotfireimages.com.) Thanks to Rich Wandschneider and Frank Conley of Fishtrap Writer's Conference. At one of their winter conferences I met Beverly Fisher, my wonderful publicist at Harcourt. Beverly mentored the work during the beginning stages as it was considered by members of Harcourt's San Diego staff.

I shall forever be indebted to John Daniel for his professional editing effort. As editor, advisor, candid critic, and ultimately good friend, John assisted in improving the manuscript to the point where it could finally be submitted.

Thanks also to Kati Steele Hesford of Harcourt, not only for her initial recognition of the book's potential, but for her skill as an editor, and her patience in putting up with me. Also to Victoria Shoemaker, my agent, for her constant support and encouragement.

I thank my mother and father. Throughout these nine years (just as during all of my life), my mother Patty has been my most constant supporter and friend. Much of what I know of courage and fortitude, I learned from her. Also to my father, Toots, who taught me that it is honorable to do good work. I also thank Eric, my son, for his encouragement, faith, and good cheer.

Finally, I thank all smokejumpers. Smokejumpers are always telling stories, each one revealing a glimpse of what our work means to us. Thank you for your stories. And thank you for creating a community in which a person like me could confirm something every man and woman needs to know—that at times, beyond all doubt, they can be brave and they can be strong.

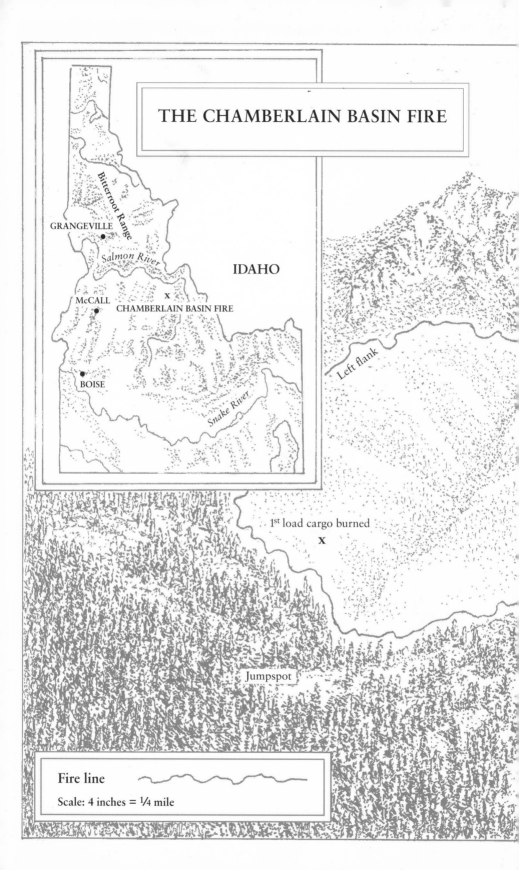

THE CHAMBERLAIN BASIN FIRE

GRANGEVILLE

Bitterroot Range

Salmon River

IDAHO

McCALL

X

CHAMBERLAIN BASIN FIRE

BOISE

Snake River

Left flank

1st load cargo burned

X

Jumpspot

Fire line

Scale: 4 inches = ¼ mile